CONSTELLATIONS

THE STORY OF SPACE TOLD THROUGH
THE 88 KNOWN STAR PATTERNS IN THE NIGHT SKY

GOVERT SCHILLING
STAR MAPS BY WIL TIRION

BLACK DOG
& LEVENTHAL
PUBLISHERS
NEW YORK

Black Dog & Leventhal Publishers
Hachette Book Group
1290 Avenue of the Americas
New York, NY 10104

www.hachettebookgroup.com
www.blackdogandleventhal.com

Image credits can be found on page 216 and constitute an extension of this copyright page.

First Edition: June 2019

Black Dog & Leventhal Publishers is an imprint of Running Press, a division of Hachette Book Group. The Black Dog & Leventhal Publishers name and logo are trademarks of Hachette Book Group, Inc.

The publisher is not responsible for websites (or their content) that are not owned by the publisher.

The Hachette Speakers Bureau provides a wide range of authors for speaking events. To find out more, go to www.HachetteSpeakersBureau.com or call (866) 376-6591.

Print book interior design by Sheila Hart Design

Library of Congress Cataloging-in-Publication Data

Names: Schilling, Govert, author. | Tirion, Wil, illustrator.
Title: Constellations : the story of space told through the 88 star patterns in the night sky / Govert Schilling ; original star maps by Wil Tirion.
Description: New York : Black Dog & Leventhal, [2019] | Includes bibliographical references and index.
Identifiers: LCCN 2018048884| ISBN 9780316483889 (hardcover : alk. paper) | ISBN 9780316483896 (ebook : alk. paper)
Subjects: LCSH: Constellations—Popular works. | Astronomy—Popular works. | Stars—Popular works.
Classification: LCC QB44.3 .S3275 2019 | DDC 523.8—dc23
LC record available at https://lccn.loc.gov/2018048884

Printed in China

1010

10 9 8 7 6 5 4 3

CONSTELLATIONS

CONTENTS

INTRODUCTION

IN 1964, ASTRONOMERS DETECTED A MYSTERIOUS SOURCE of X-rays in the Milky Way. Subsequent research revealed that the energetic radiation is produced by hot gas, ripped off a giant star by the fierce gravity of an orbiting black hole. In fact, this was the first black hole ever identified in the universe, at a distance of some 6,000 light-years.

The black hole is officially known as Cygnus X-1, which is weird. *Cygnus* is Latin for "swan," but whatever the remote black hole is devouring, it isn't birds. So why would an astronomical object be named after a waterfowl?

Cygnus X-1 is not alone. Some 54 million light-years away—well beyond our own Milky Way galaxy—is a large swarm of other galaxies. It is known as the Virgo Cluster. *Virgo* is Latin for "virgin"—quite a remarkable name for a collection of two thousand or so Milky Ways. Yet another interesting astronomical object—a galaxy emitting radio waves—is catalogued as Centaurus A, but don't expect to find a half man half horse galloping around there.

These funny and archaic names, which often show up in contemporary scientific publications like the *Astrophysical Journal*, refer to the part of the sky in which the objects are located. In prehistoric times, early sky watchers noted a conspicuous group of stars in the northern summer sky that vaguely resembles the shape of a flying swan. Consequently, they called this part of the night sky the Swan, or Cygnus. And since the X-ray-emitting black hole was the first one to be observed in this same area of the night sky—in the same *constellation*—it earned the official designation Cygnus X-1. Likewise, the galaxy cluster mentioned above is located in a part of the sky known as Virgo, the Virgin, and the radio galaxy Centaurus A can be found in the constellation that was named after the mythological creature thousands of years ago.

I have always enjoyed the fact that twenty-first-century scientists who study black holes, dark matter, and galaxy collisions still use these age-old names and concepts to denote their objects of research. That Hercules, Orion, and Pegasus—names that were very familiar to the ancients—still feature on the pages of magazines like *Nature Astronomy* and *Science*. It underscores the relation of today's scientists to their predecessors of many centuries ago. After all, just like the Babylonians, Egyptians, and Greeks, we are still trying to make sense of the universe we live in.

Moreover, it reminds us of the fact that every single object in the universe, and everything that happens in space, is observed within the confines of one constellation or the other. In 1054 AD, a star exploded in the constellation Taurus, the Bull. In 1930, Pluto was discovered in the constellation Gemini, the Twins. In 2015, astronomers detected gravitational waves from two colliding neutron stars, 130 million light-years away in a galaxy in Eridanus. And half a century ago, Neil Armstrong's famous "giant leap for mankind" took place when the moon was in Virgo.

Thus, the eighty-eight constellations of the night sky offer eighty-eight windows on the entire universe and on the history of astronomy. Each new insight in the contents and evolution of the cosmos was gained by astronomers training their instruments to a particular point on the sky, and every spacecraft encounter, comet discovery, or Mars landing took place within the borders of one of the eighty-eight constellations.

Often enough, astronomy websites and newspaper stories report on exciting new discoveries and events, like an extremely remote galaxy at the edge of the observable universe, a monstrous black hole, or a space probe flyby of a distant planet. This book uniquely ties all these finds and occurrences, both historical and more recent, to the constellation in which they happened.

My hope is that after you read and browse through this book, the night sky will never be the same again. Right above your head, in familiar constellations like Orion or Scorpius, or in obscure ones like Camelopardalis and Norma, is where the science of astronomy has matured and where milestone discoveries in the study of the universe have played themselves out.

I thank the team at Black Dog & Leventhal Publishers for their confidence in this exciting project, and I am honored that world-famous astrocartographer (and fellow Dutchman) Wil Tirion was willing to draw the beautiful constellation maps, which have been specially designed and adapted for this book. Most of all, however, I have to thank the astronomers of the past and the present for their perseverance in studying the wider universe we are all part of, and for presenting us with a wealth of mind-boggling discoveries and breathtaking images.

—**GOVERT SCHILLING**

◀ **The small constellation Corvus (the Raven) in a color edition of Johann Bayer's 1603 star atlas *Uranometria*.**

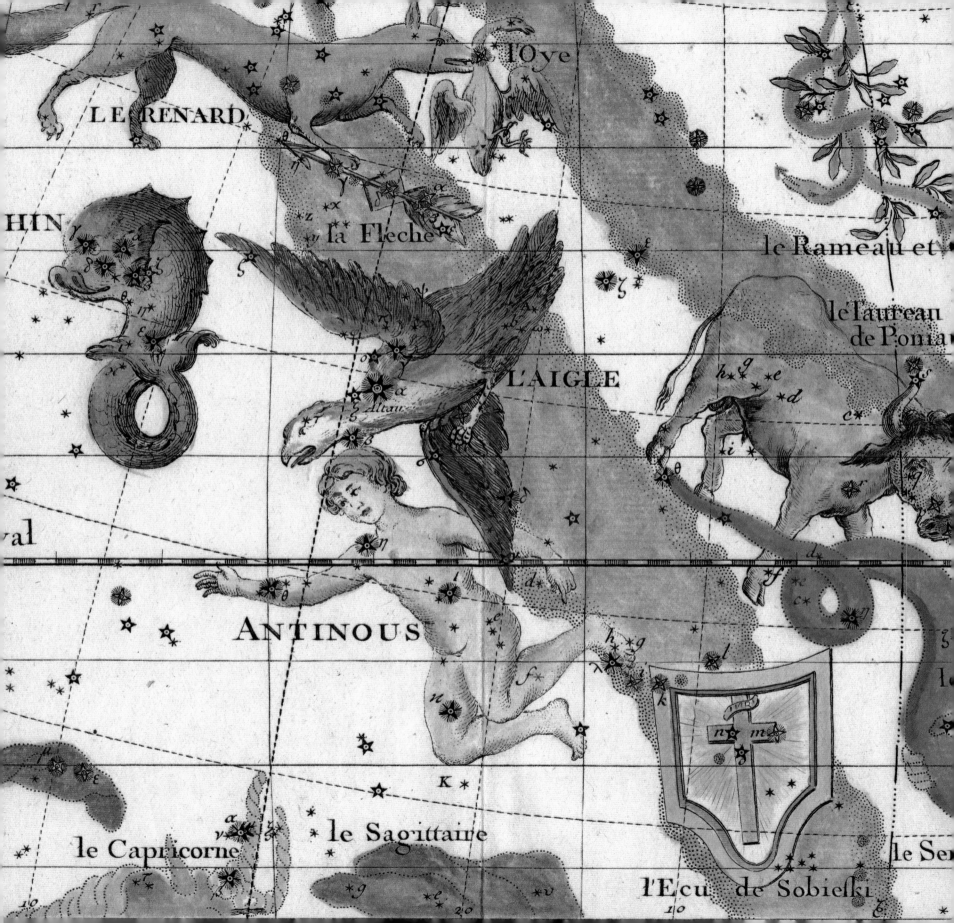

CONSTELLATIONS THROUGH THE AGES

ON THE AFRICAN SAVANNAH, hundreds of thousands of years ago, early hominids must have marveled at the night sky. No one could fathom the true nature of the stars, but seeing the same patterns slowly revolving around the Earth night after night and year afte r year must surely have fired the imagination. No doubt that the most conspicuous of these patterns received names as soon as *Homo sapiens* started to develop speech. Names that were related to gods and demons, to humans and animals, to life and death.

The origin of some of our present constellations is lost in the mists of time. The Ox (or Bull), the Giant (or Hunter), the Lion, and the Bear—they probably predate the Sumerian invention of cuneiform, one of the oldest forms of writing. Many of these old constellation names, as well as the names of sixty-six individual stars, appear in the Babylonian MUL.APIN, a famous pair of Mesopotamian cuneiform tablets dating back to approximately 1000 BC.

Around the fourth century BC, most of the ancient Mesopotamian constellations were introduced to ancient Greece and Egypt. A beautiful relief on the ceiling of the Egyptian temple of Dendera, dating back to the year 50 BC or so, shows many well-known constellations like the Bull, the Twins, and the Lion, next to ancient Egyptian ones like the Hippo and the Ox's Foreleg. (By the way, the relief that is now in Dendera is a copy; the original Zodiac of Dendera, as it is called, is in the Louvre Museum in Paris.)

The basis for our present-day constellations was laid by Greek astronomer Claudius Ptolemy, who lived and worked in Alexandria in the second century AD. In his *Mathēmatikē Syntaxis*—a Greek collection of thirteen books on mathematics and astronomy—he listed forty-eight constellations that are still recognized today, including the twelve constellations of the zodiac. Ptolemy also included a catalog of 1,028 stars, based on an earlier list compiled by Hipparchus of Nicaea some 250 years earlier.

In the ninth century AD, the Greek constellations were adopted by Islamic astronomers, who translated Ptolemy's work into Arabic as *al-Majisṭī* (The Greatest). However, they introduced their own proper names for many of the bright stars in the night sky, most of which are

◀ **This map from a 1795 star atlas shows the now-obsolete constellations Antinous (center) and Taurus Poniatovii (Poniatowski's Bull, right).**

still in use today. Star names like Betelgeuse, Aldebaran, Deneb, and Rasalhague are all derived from ancient Arabic names. Both the Greek constellations and the Arabic names were subsequently introduced to early European scholars when Ptolemy's book was translated from Arabic into Latin (as the *Almagest*) in the late twelfth century.

Obviously, other cultures around the world who had never been in contact with the civilizations of the Middle East—like the Aztecs, Mayas, and Incas in the Americas, the Aboriginals in Australia, the Polynesians in the Pacific, and of course the Chinese—had their own constellations. Ancient Chinese star maps reveal a bewildering number of small constellations or "asterisms" (small groups of stars), although Chinese astrologers also recognized the Big Dipper as one pattern, called Bei Dou (Northern Ladle). Meanwhile, both in South America and in Australia, the sinuous dark clouds in the star-studded band of the Milky Way were seen as "dark constellations," such as the Incan Partridge, the Llama (and the Baby Llama), the Serpent, and the Aboriginal Giant Emu in the Sky.

From Peru and Australia, all of the southern sky is visible, including the Southern Cross, the bright center of the Milky Way, and the two Magellanic Clouds, which are small satellite galaxies of the Milky Way. However, Renaissance scientists in Europe knew next to nothing about the stars of the southern hemisphere. As a result, European maps of the southern sky—like the famous 1515 woodcuts of Albrecht Dürer—are notoriously incomplete.

That all changed when intrepid explorers started to sail their ships over the seven oceans. In 1595, almost 250 Dutch seafarers and navigators embarked on a two-year trip around Africa and to the East Indies in a small fleet of four ships. Among them were Pieter Dirkszoon Keyser and Frederick de Houtman, who had been trained in astronomy by Flemish cartographer Petrus Plancius. Keyser and de Houtman mapped the southern sky and introduced twelve new constellations, which they named after the miraculous creatures they encountered during their trip, like Volans (the Flying Fish), Apus (the Bird of Paradise), and even Indus (the Indian).

In 1598, Plancius's twelve new constellations (as well as Columba, the Dove, which he already had "invented" in 1592, and Coma Berenices Berenice's Hair, which had been introduced in 1536 by Caspar Vopel) were immortalized on a celestial globe by Jodocus Hondius, and they were

◀ In 1627, Julius Schiller published
his christianized view of the heavens.

▶ Part of the Egyptian
"Zodiac of Dendera,"
with the constellation
Pisces (the Fishes) on
the upper left.

adopted by German astronomer Johann Bayer in his beautifully illustrated 1603 star atlas *Uranometria*. In 1612, Pl ancius invented another
eight constellations, but of those, only Camelopardalis (the Giraffe) and
Monoceros (the Unicorn) are still officially recognized. (Note that many
of the historical star atlases show the constellations mirror-reversed, as if
we are viewing the celestial sphere from the "outside.")

Not everyone was happy with the fact that the celestial sky was
slowly turning into a tapestry of pagan gods, mythological heroes, and
strange animals. In 1627, German astronomer and devout Christian
Julius Schiller published his *Coelum Stellatum Christianum* (Christian
Starry Heavens), in which he had turned the twelve constellations of
the zodiac into the twelve apostles of Christ, Cygnus the Swan into
the cross of Golgotha, and the Big Dipper into the fishing boat of
St. Peter, to give just a few examples. Although the illustrations in
Schiller's atlas are among the most beautiful in celestial cartography,
his plan to Christianize the constellations never became popular.

However, new constellations kept being added to the existing list,
in particular by Polish astronomer Johannes Hevelius (in 1687) and by
French astronomer Nicolas Louis de Lacaille (in the 1750s). Hevelius
introduced ten new constellations, seven of which are still in use, including Canes Venatici (the Hunting Dogs) and Lynx. Lacaille came up with
fourteen new (but rather inconspicuous) constellations in the southern
sky, which he named after scientific instruments, like Fornax (the Oven),
Telescopium (the Telescope), and Antlia (the Air Pump). Lacaille also
divided the very large Ptolemaic constellation Argo Navis (Ship *Argo*)
into three smaller ones. The Latin names of Lacaille's new inventions first
appeared in his 1763 publication *Coelum Australe Stelliferum*; for that
reason, 1763 will be considered the official year of introduction of these
constellations throughout this book.

In eighteenth- and nineteenth-century star atlases, astronomers
sometimes tried to add new constellations of their own, but none of
these have survived. Thus, in the early twentieth century, a grand total
of eighty-eight constellations were "officially" in use. However, their
boundaries on the sky had never been precisely specified. Sure, every
naked-eye star definitely belonged to a particular constellation—they

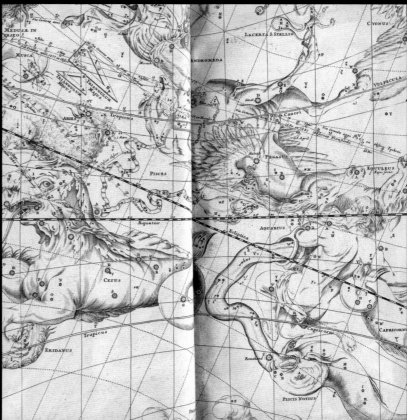

▲ Aries (the Ram), Cetus (the Whale), Pegasus, and Aquarius (the Water Bearer)
are some of the constellations visible on this map from Johann Gabriel Doppelmayr's
1742 *Atlas Coelestis*.

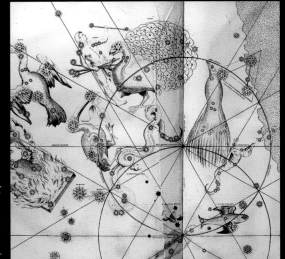

◄ Johann Bayer's 1603 star atlas *Uranometria* was the first one to depict the new southern constellations that were introduced by Dutch seafarers Pieter Dirkszoon Keyser and Frederick de Houtman.

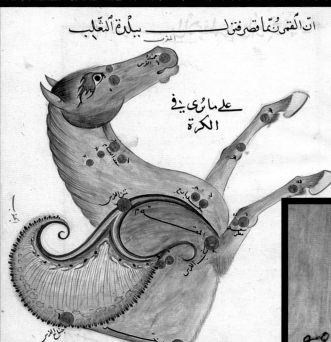

▲ The flying horse Pegasus features prominently in a fifteenth-century Arabic star atlas that was based on Abd al-Rahman al-Sufi's *Book of Fixed Stars*.

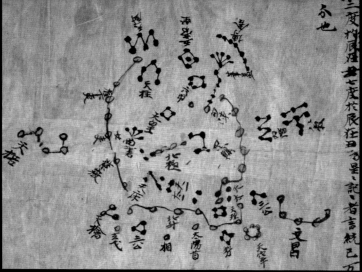

◄ Amid numerous smaller groups, the characteristic shape of the Big Dipper can be recognized on the seventh-century Chinese Dunhuang star chart (bottom).

had been given a Greek letter designation by Johann Bayer according to their brightness, or a number designation by English astronomer John Flamsteed according to their position. But for many faint stars and nebulae in the border regions between two constellations, it was often not evident to which one they belonged.

In 1930, Belgian astronomer Eugène Delporte finally drew up the modern boundaries between the constellations, which were subsequently approved by the International Astronomical Union. From then on, every object in the universe, be it a nearby dwarf star or a remote galaxy, belongs to one and only one constellation, and every astronomical discovery or planetary exploration event occurred within the borders of one of those eighty-eight official patterns in the sky. That realization is the basis of this book.

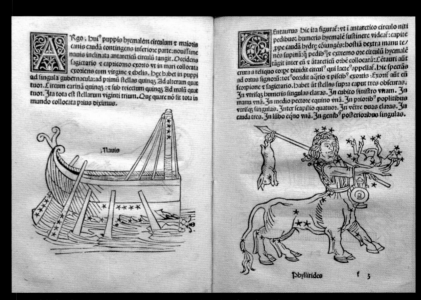

▲ Published in 1482, *Poeticon Astronomicon* is a book with constellation drawings and mythological stories, based on the classical work by Latin author Gaius Julius Hyginus.

▲ One of the many beautiful illustrations in Andreas Cellarius's *Harmonica Macrocosmica* (1660) shows the zodiacal band with its twelve constellations encircling the Earth.

TWELVE OF THE ORIGINAL FORTY-EIGHT CONSTELLATIONS that were described by Claudius Ptolemy in his *Almagest* are special: the twelve constellations of the zodiac. Many people know their names: Aries (the Ram), Taurus (the Bull), Gemini (the Twins), Cancer (the Crab), Leo (the Lion), Virgo (the Virgin), Libra (the Scales), Scorpius (the Scorpion), Sagittarius (the Archer), Capricornus (the Sea Goat), Aquarius (the Water Bearer), and Pisces (the Fishes). So what's so unique about them?

This becomes clear when we realize that the Earth and the other planets in the solar system all orbit the sun essentially in a flat plane. Even the moon's orbit around the Earth is only tilted a few degrees with respect to this plane. As a result, from our vantage point on Earth, we observe the sun, the moon, and the other planets always in a relatively narrow band of sky around us. This band is called the zodiac, and it coincides more or less with these twelve ancient constellations.

In the course of a year, the sun appears to move through the constellations of the zodiac along the so-called ecliptic. The ecliptic is in fact the projection of Earth's orbit on the celestial sky. The moon and the planets are never too far away from the ecliptic, so they, too, can usually be found in one of the constellations of the zodiac. In other words: Jupiter can be seen in Leo or Sagittarius, but never in Ursa Major (the Great Bear), because that constellation does not belong to the zodiac. Likewise, the moon may form a nice pair on the sky with the star Antares in Scorpius, or with Aldebaran in Taurus, but it will never be found in Cassiopeia.

More than two thousand years ago, the zodiac (and the ecliptic) was divided into twelve equally sized parts, each measuring 30 degrees on the sky. These segments were called the zodiacal signs, and they received the same names as the constellations with which they more or less coincided. On March 20 or 21, at the so-called vernal equinox, the sun enters the sign of Aries; about a month later, when it has moved 30 degrees along the ecliptic, it enters the sign of Taurus, and so on. This is the basis of your "star sign"; if you're born on November 30, for example, the sun is in the sign of Sagittarius on your birthday, and you are said to be a Sagittarius. Likewise, if you're born on July 4, you're a Cancer.

Some people believe that your star sign (as well as the sky positions of the moon and the planets at the time of your birth) determine your

▲ The zodiacal constellation Gemini (the Twins) as it appears in the *Leiden Aratea*, an illuminated ninth-century manuscript based on the *Phaenomena* by Aratus.

▶ In the German star atlas *Uranographia* by Johann Elert Bode, the inconspicuous zodiacal constellation Cancer (the Crab) is labeled as Krebs.

personality, but that's plain old-fashioned superstition. Moreover, because of precession—the very slow periodic change in the orientation of the Earth's axis—the zodiacal signs no longer correspond to the constellations: on July 4, the sun is actually in the constellation Gemini (the Twins), although it is still said to be in the zodiacal sign of Cancer by astrologers. (The ecliptic even passes through a thirteenth constellation, Ophiuchus, which isn't part of the original zodiac at all.)

Pseudoscience aside, the twelve constellations of the zodiac (plus Ophiuchus) are very important in astronomy, especially for amateur astronomers observing the night sky. After all, this is where most of the solar system action occurs, including eye-catching conjunctions of the moon and the planets, and lunar and solar eclipses. New planets—as well as many asteroids—have also been discovered in constellations of the zodiac, including Uranus in Taurus (the Bull), Neptune in Aquarius (the Water Bearer), and Pluto in Gemini (the Twins). Finally, to communicate with planetary probes orbiting other planets, or with landers and rovers on Mars, space scientists always have to aim their radio dishes somewhere along the zodiac. Little wonder then that these twelve constellations play an important role in this book.

▶ Sagittarius (the Archer) draws his bow on a sixteenth-century celestial globe by German-Flemish cartographer Gerardus Mercator.

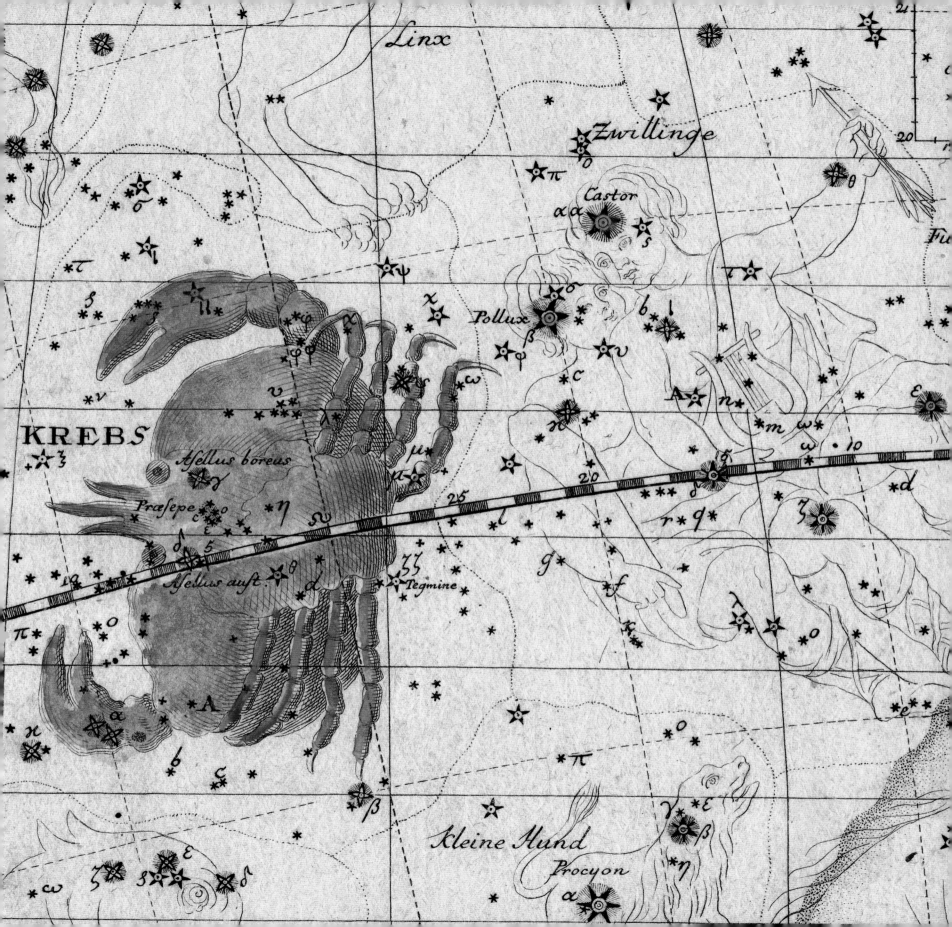

AN ASTRONOMY PRIMER

IN ONE SENSE, ASTRONOMY IS THE OLDEST SCIENCE. After all, our earliest ancestors must already have wondered about the motions of the sun and the moon, and about the true nature of the stars. In another sense, however, astronomy is a very *young* science; almost all our current knowledge about the universe has been acquired over the past four centuries, after the invention of the telescope.

Ancient Greek astronomers believed the Earth to be the center of the universe. This was considered common wisdom until 1543, when Polish astronomer Nicolaus Copernicus published his heliocentric world view, with the Earth orbiting the sun instead of the other way around. It took another couple of centuries until astronomers reliably determined the distances to other stars, while the discovery of the true nature of spiral nebulae (galaxies like our own Milky Way) and of the expansion of the universe is less than a century old.

We now know that Earth is just one of eight planets orbiting the sun, and one of the smaller ones at that—Jupiter and Saturn are much larger. Our sun is a run-of-the-mill star, very much comparable to the thousands of stars that can be seen in the night sky—a giant sphere of hot gas (mainly hydrogen and helium), held together by gravity, and producing light and heat through spontaneous nuclear fusion reactions in its interior.

Together with a few hundred billion other stars and countless star clusters and nebulae, the sun makes up our Milky Way galaxy, a flattened disk, more than 100,000 light-years across. (One light-year is the distance a ray of light, traveling at 186,282 miles per second, covers in one year: some 5.88 trillion miles.) Our Milky Way, in turn, is just one of approximately a hundred billion other galaxies in the entire observable universe.

Over the past century, astronomers have not only discovered our truly insignificant place in space, they have also unraveled cosmic history, even though the details of the universe's birth (known as the big bang) are still eluding them. There is no doubt, however, that 13.8 billion years ago all of space was incredibly more compact and filled with a seething brew of elementary particles, including mysterious particles of dark matter.

While space itself expanded dramatically over time, diluting matter and energy in the process, gravity caused tiny overdensities to grow into ever-larger clumps of dark matter, which subsequently also attracted normal atoms. Relatively cool clouds of gas condensed into the first building blocks of today's galaxies, and deep within these clouds, the first stars ignited. Collisions and mergers of these irregular protogalaxies produced the giant spiral and elliptical galaxies we see all around us, distributed in small groups and larger clusters and superclusters.

▼ In 1992, NASA's Galileo spacecraft captured this image of the Earth and the moon from a distance of 6 million kilometers.

As soon as the first stars formed, from gravitationally collapsing cloud cores, the chemical composition of the universe started to change, albeit slowly and minutely. Under the extreme circumstances in stellar interiors, atoms of hydrogen and helium started to fuse into other elements, including carbon and oxygen. In the most massive stars, even heavier atoms were formed, like neon, silicon, and iron.

At the end of their lives, low-mass stars like our own sun shed their outer layers into space as glowing planetary nebulae, while their cores contract into extremely dense white dwarf stars. More massive stars end

◀ In northern Chile, European astronomers are building the Extremely Large Telescope, the largest optical telescope in history.

At the end of their lives, stars like our Sun blow their outer layers into space, creating colorful expanding shells of gas known as planetary nebulae.

Our solar system was born 4.6 billion years ago in a protoplanetary disk of gas and dust. Similar disks have been found around many other stars.

their lives in catastrophic supernova explosions. When it runs out of nuclear fuel the star is blown apart, leaving nothing more than a hyper-compact neutron star, or even a black hole, while the supernova inferno produces heavy elements like gold, lead, and uranium. What's important here is that stellar winds, planetary nebulae, and supernova explosions dump new elements into interstellar space. As a result, the universe around us no longer consists solely of hydrogen and helium, like it did 13.8 billion years ago, but it contains about 1 percent heavier elements.

Over time, these heavier atoms, some of them locked up in carbon-bearing molecules, ended up in dusty stellar nurseries that hatched new generations of stars. And although the relative abundance of the heavier elements was rather small, the absolute amounts of iron, nickel, carbon, silicon, oxygen, manganese, aluminum, uranium, and so on were more than sufficient for the formation of protoplanetary systems—flattened, rotating disks of gas and dust around newborn stars.

Within these disks, dust particles coalesced into pebbles, pebbles stuck together to form "planetesimals," and planetesimals accreted into protoplanets. From the violent collisions of protoplanets,

▲ Thanks to its majestic ring system, Saturn is one of the most photogenic planets in our solar system. This photo was made by the Hubble Space Telescope.

full-fledged worlds like our own Earth were born. Organic molecules rained down into lukewarm pools of surface water, the first living cells emerged, and the rest is history. A few billion years later, *Homo sapiens* looks up in wonder at the sky, starts to ask questions and to build telescopes, and discovers its place in space and time.

The invention of the telescope, just more than four hundred years ago, marked the beginning of an ongoing quest to design and build more sensitive instruments to study the universe that we are so intimately part of. Astronomers have discovered ways to observe the invisible parts of the electromagnetic spectrum, using infrared cameras, radio telescopes, and X-ray detectors. They put their equipment into Earth orbit to get rid of the blurring and absorbing effects of the atmosphere. They send space probes to explore planets and moons in our own solar system, and they use giant computers to sift through the data gathered by robotic telescopes.

The pace of new discoveries has steadily increased over the decades: dozens of planetary satellites, hundreds of comets, tens of thousands of asteroids. Stellar binaries, variable stars, red giants, white dwarfs, nebulae, and supernova remnants have been discovered, including pulsars, rapidly spinning neutron stars that emit pulses of radio waves; and quasars, remote galaxies with extremely luminous cores, fueled by massive black holes; the soft radio hiss that is all that remains of the energy of the big bang; thousands of exoplanets orbiting stars other than the sun; powerful gamma-ray bursts; mysterious brief explosions known as fast radio bursts; and gravitational waves, ripples in the very fabric of space-time.

Many questions have been answered, but enough riddles remain. We still don't know whether life is rare or common in the universe. We have no clue about the true nature of dark matter, let alone about dark energy, the mysterious force that is accelerating the expansion of the universe. What happened during the big bang is just as mysterious as what goes on inside black holes, and new theories are required for a deeper understanding. We don't even know if our universe is unique or part of an infinite multiverse.

But just like the scientists of centuries past, today's astronomers know that there's only one way to make progress: never give up watching the skies. The heavens above us, once the domain of deities and dragons, has turned into a vast celestial stage on which the evolution of the universe is playing itself out right before our eyes. Peering through the eighty-eight windows of the constellations with everything we have at our disposal, we are trying not to miss a single detail of the great cosmic drama.

▲ Nicknamed the Sunflower Galaxy, M63 is a beautiful spiral in the constellation Canes Venatici (the Hunting Dogs), comparable in size to our own Milky Way.

▶ Dark sunspots are caused by magnetic fields, and often accompanied by bright flares. Our sun is just one of a few hundred billion stars in our Milky Way galaxy.

WATCHING THE SKIES

WE ALL KNOW THAT THE SUN rises in the east and sets in the west. In the past, people thought that the sun really orbits Earth. We now realize that the daily motion of our star is an apparent motion, caused by the fact that Earth rotates around its axis. For the same reason, the stars in the night sky display the same diurnal motion, rising in the east and setting in the west.

If you were standing on the geographical North Pole of the rotating Earth, the stars would appear to move around you horizontally, without rising or setting. The only point on the sky that would not appear to move is the point right above your head (in the zenith). Instead, all celestial bodies would appear to rotate around this point. Just by coincidence, a pretty bright star is located very close to the celestial pole: the Pole Star, in the constellation Ursa Minor (the Little Bear).

For an observer on Earth's equator, the stars would move around vertically. The Pole Star would sit on the northern horizon, while the south celestial pole would be directly opposite, on the southern horizon. (Unfortunately, there is no bright star close to the south celestial pole.)

Somewhere between the North Pole and the equator, stars appear to move diagonally, and the Pole Star is located somewhere between the zenith and the northern horizon. But Earth's axis still points in the direction of the Pole Star, so even for observers in North America or Europe, the whole sky appears to rotate around this star. Stars and constellations that are relatively close to the Pole Star (like the Big Dipper, Cassiopeia, and Cepheus) never set, and can be seen at any time of night throughout the year, at least on cloudless nights. These constellations are known as *circumpolar*. Then again, for an observer in North America or Europe, the south celestial pole is always below the horizon, and the constellations in this part of the sky, like the Southern Cross, can *never* be seen.

In between are the constellations that do rise and set, like Orion, Virgo, or Aquila. Sometimes they can be seen only in the early hours of the night, before they disappear below the western horizon. At other times they rise in the east only around midnight. Sometimes they are visible throughout the night, from sunset to sunrise; at other times you can't see them at all. This all depends on Earth's yearly motion around the sun, which ensures that each season has its own characteristic constellations.

The planets, which also orbit the sun, are always found in or near one of the constellations of the zodiac, as has been explained before. Planets basically look like bright stars, but they don't have a fixed position among the stars. Instead, they slowly wander across the sky in funny loops, caused by the fact that we observe them from a moving platform—our own Earth. Just like the moon, the planets do not produce light by themselves; they just reflect light from the sun. (In fact, a pair of binoculars or a small telescope already reveals the phases of Venus, which are very

▶ A beautiful conjunction of the moon and the planets Venus and Jupiter, seen above the enclosures of the European Very Large Telescope in northern Chile.

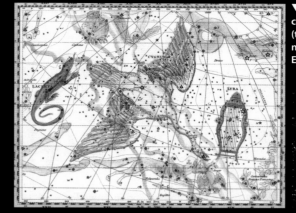

▶ **Cygnus (the Swan), as depicted in Alexander Jamieson's 1822 *Celestial Atlas*. Lacerta (the Lizard) and Lyra (the Lyre) are to the Swan's left and right, respectively.**

▼ **The small but prominent constellation Crux Australis (the Southern Cross) can never be seen from most of Europe and North America.**

similar to the phases of the moon and are caused by the varying illumination from the sun.)

With the naked eye, you can see five planets: Mercury, Venus, Mars, Jupiter, and Saturn, although it is very rare for all five of them to be visible during one single night. If you're far away from the light pollution of cities and towns, you will also be able to see a few thousand stars, not just the pretty bright ones that define the constellations, but also countless, fainter ones. Some stars can be seen to be double stars (or binaries), like the middle star in the handle of the Big Dipper. Others have a quite conspicuous color, like Aldebaran (in the constellation Taurus, the Bull) and Betelgeuse (in Orion), which are both orange.

Sometimes a meteor, or "shooting star," appears unexpectedly: a streak of light high up in the sky, caused by a cosmic pebble that burns up in our atmosphere. With some patience, you're also bound to see one or more artificial satellites—faint "stars" that slowly move across the sky in a straight line. If you're lucky (or if you use a smartphone app to give you advance warning), you may be able to see the very bright International Space Station passing overhead.

Around the time of the new moon, the hazy band of the Milky Way can be seen stretching itself across the sky from horizon to horizon. This is in fact the smeared-out light of hundreds of thousands of faint stars in our own Milky Way galaxy, which we observe from inside because our sun is also part of it.

To really enjoy the night sky, you may want to use a pair of binoculars (preferably mounted on a tripod) or a small telescope. These optical aids will reveal many more binary stars, as well as dozens of star clusters, nebulae, and galaxies. Many of them can be found in the constellation maps in this book, and with just a little practice you will be able to find most of them.

Astronomy is a great hobby. If this book ignites your enthusiasm, lend or buy other introductory astronomy books, subscribe to an astronomy magazine, visit a star party, public observatory, or planetarium, become part of an astronomy club, buy a telescope, and enjoy the many astronomy-related websites and blogs on the internet. It's a vast and magnificent universe out there.

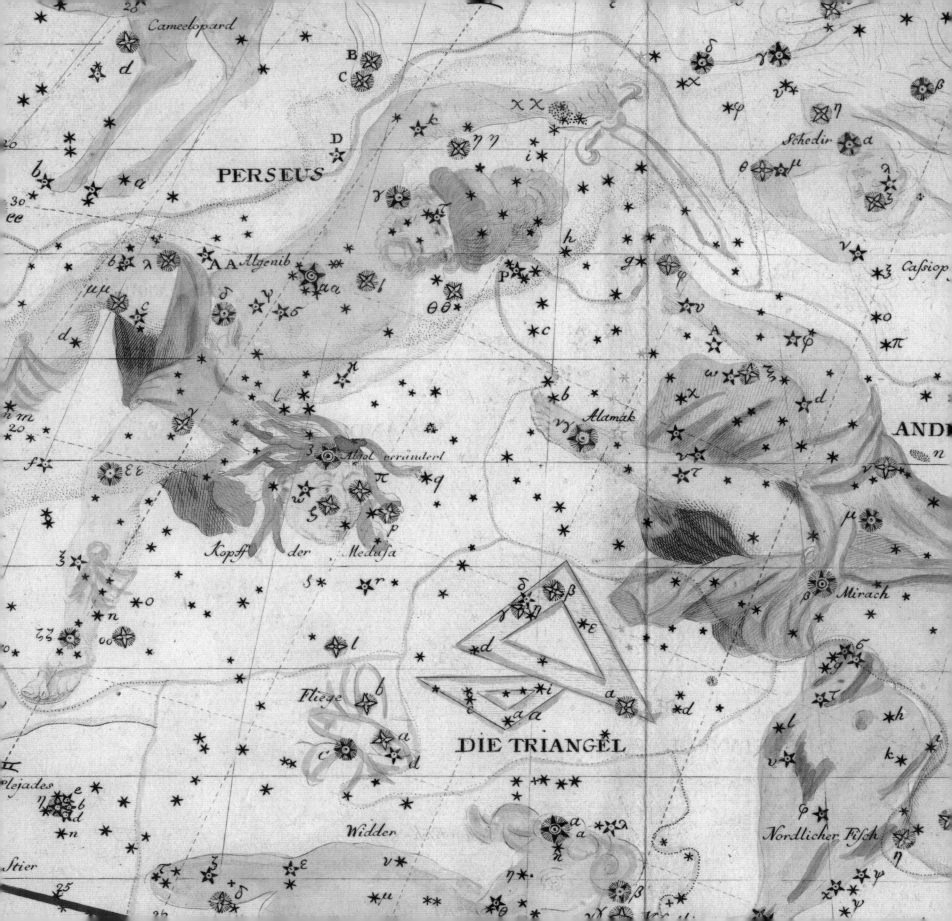

HOW THIS BOOK IS ORGANIZED

THE CONSTELLATIONS IN THIS BOOK are arranged in the alphabetical order of their official Latin names. The Passport section on the top of the first page contains factual information about the constellation, starting with the official Latin name and the (unofficial) English name. "Genitive" denotes the way objects in this constellation are designated. For instance, the star Alpha (α) in the constellation Cygnus is known as α Cygni. Next is the official three-letter abbreviation of the constellation name, according to the International Astronomical Union (IAU).

"Origin" refers to the person who first introduced the constellation (see also page 13). (Note that most of the constellations "introduced" by Claudius Ptolemy are actually much older; some of them date back to Sumerian times.) "Area" indicates the size of the constellation in the sky, measured in square degrees (a degree is about twice the apparent diameter of the full moon). The number of naked-eye stars in the constellation is the number of stars that someone with extremely good eyesight may be able to see without optical aid on an extremely clear night from a very dark place; in general, the number of stars you will be able to see is much smaller.

The Passport section also gives the names of the bordering constellations, which eventually will help you to get more familiar with their relative positions in the sky. Finally, "Best visibility" tells you in which months a particular constellation reaches its highest altitude above the horizon around midnight, and from which areas on Earth you will be able to see the *whole* constellation. A few months earlier, the constellation will reach its highest point in the sky before midnight; a few months later, it will reach its highest point only after midnight. The color of the Passport section is related to the constellation's location on the sky: green for northern hemisphere constellations, yellow for equatorial constellations, and blue for southern hemisphere constellations.

The constellation maps are all drawn to the same scale. They show the majority of naked-eye stars, plus a selection of star clusters, nebulae, and galaxies, most of which can be seen only with binoculars or a small

◄ **Perseus rescues Andromeda on this map from a color edition of Johann Bode's *Uranographia*.**

telescope. The size of a star symbol is a measure of its apparent brightness, which for stars is given in *magnitudes*, with magnitude 0 for the brightest ones and magnitude 6 for the stars that are only just visible to the naked eye (see the legend on the opposite page). Some stars are denoted by their official IAU name and/or their Bayer designation: a Greek letter, where α is usually the brightest star in the constellation, β the second brightest, and so on (see the Greek alphabet on this page). Note that stars can also bear different designations, like numbers, letters, or various catalog entries.

The grid of blue lines on the constellation maps mark the celestial coordinate system of *right ascension* and *declination*. Right ascension is comparable to geographical longitude on Earth, but is measured in time units instead of degrees, from 0^h to 24^h. Declination is measured in degrees north or south (+ or –), just like geographical latitude on Earth. The yellow lines are the official IAU borders of the constellation. Originally, these borders were parallel to the celestial coordinate grid, but as a result of precession (the slow change in the orientation of the Earth's axis), they are a little bit skewed today. Where applicable, the ecliptic (see page 19) and the Milky Way are also shown on the maps.

Various symbols are used to denote so-called deep-sky objects on the sky (see legend). Open star clusters are young groups of tens or hundreds of stars that are born together, usually in the spiral arms of our Milky Way galaxy. The brightest and most famous one is the Pleiades star cluster in the constellation Taurus (the Bull). Globular star clusters are huge spherical collections of hundreds of thousands of very old stars. They are distributed in a huge, more-or-less spherical halo around the center of our Milky Way galaxy,

Nebulae are glowing clouds of hot interstellar gas in which new stars are born. Dark clouds contain a lot of cold dust and can be seen if silhouetted against a bright nebulous background. Planetary nebulae are the expanding shells of gas that are blown into space by dying sun-like stars. Galaxies are huge collections of billions of stars, comparable to our own Milky Way galaxy; they are often grouped in huge galaxy clusters.

Many star clusters, nebulae, and galaxies are designated by a number from the *New General Catalogue*, compiled in 1888 by John Dreyer. The letters NGC are left out of the maps, so NGC 2281—an open star cluster in the constellation Auriga (the Charioteer)—is just denoted by its number, 2281. Just over a hundred deep-sky objects have Messier numbers, like M27 or M81—they are listed in Charles Messier's eighteenth-century *Catalogue des Nébuleuses et des Amas d'Étoiles* (Catalogue of Nebulae and Star Clusters). A number of deep-sky objects also have (unofficial) proper names.

The Timeline section for each constellation provides an overview of significant discoveries and events that took place within the constellation's borders. Every object mentioned in the Timeline—even extremely faint dwarf stars, remote quasars, and mysterious radio bursts—can be found back on the constellation map. Timeline events like the discovery of an asteroid, the crash of a comet into Jupiter's atmosphere, or the landing of a spacecraft on the surface of Mars are preceded by a small number (❶, ❷, ❸); the same number can be found on the constellation map at the corresponding position in the sky.

For many Timeline items, an accompanying photo or illustration provides you with additional visual support. The best way to enjoy the book is to find the match between an illustration, a Timeline item, and the corresponding symbol on the constellation map. Try to imagine that this celestial position—a point in the sky that can easily be located in the constellation above your head—is where an astronomer pointed his telescope, satellite instrument, or radio dish to discover yet another astronomical gem or to witness an exciting event in the exploration of space.

Finally, the maps on pages 35 and 37 show the whole northern and southern sky and can be used to find out where a particular constellation is located with respect to its larger surroundings.

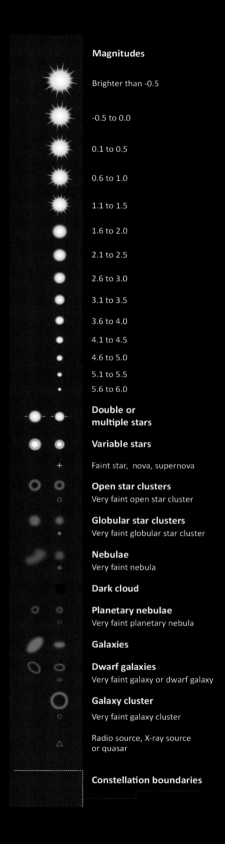

LEGEND

Magnitudes

Brighter than -0.5

-0.5 to 0.0

0.1 to 0.5

0.6 to 1.0

1.1 to 1.5

1.6 to 2.0

2.1 to 2.5

2.6 to 3.0

3.1 to 3.5

3.6 to 4.0

4.1 to 4.5

4.6 to 5.0

5.1 to 5.5

5.6 to 6.0

Double or multiple stars

Variable stars

Faint star, nova, supernova

Open star clusters
Very faint open star cluster

Globular star clusters
Very faint globular star cluster

Nebulae
Very faint nebula

Dark cloud

Planetary nebulae
Very faint planetary nebula

Galaxies

Dwarf galaxies
Very faint galaxy or dwarf galaxy

Galaxy cluster
Very faint galaxy cluster

Radio source, X-ray source or quasar

Constellation boundaries

MAP OF THE NORTHERN SKY

AQUARIUS

PISCES

EQUULEUS

PEGASUS

CETUS

DELPHINUS

ARIES

TRIANGULUM

LACERTA

ANDROMEDA

SAGITTA

ERIDANUS

VULPECULA

PERSEUS

TAURUS

AQUILA

CYGNUS

CASSIOPEIA

SERPENS
CAUDA

ORION

CEPHEUS

LYRA

CAMELOPARDALIS

AURIGA

DRACO

*North
Celestial Pole*

OPHIUCHUS

URSA MINOR

LYNX

MONOCEROS

HERCULES

GEMINI

CORONA
BOREALIS

BOÖTES

URSA MAJOR

CANCER

CANIS
MINOR

CANES VENATICI

SERPENS CAPUT

LEO MINOR

HYDRA

COMA BERENICES

LEO

ECLIPTIC

SEXTANS

VIRGO

EQUATOR

MAP OF THE SOUTHERN SKY

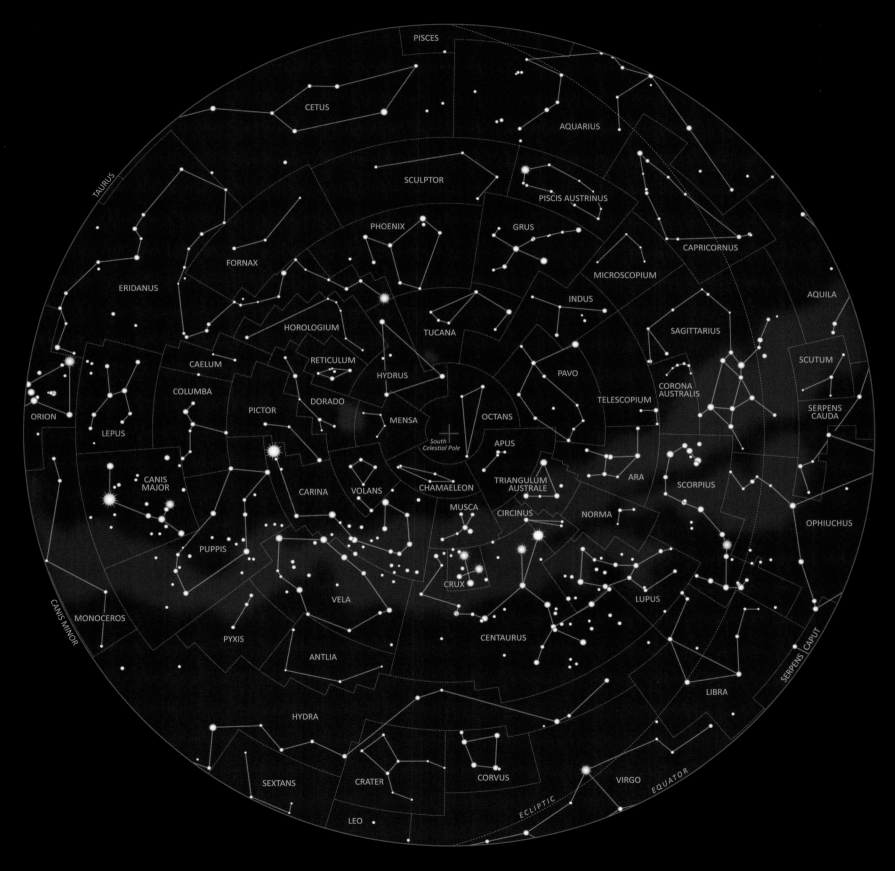

PISCES

CETUS

TAURUS

ERIDANUS

FORNAX

AQUARIUS

SCULPTOR

PISCIS AUSTRINUS

PHOENIX

GRUS

CAPRICORNUS

MICROSCOPIUM

INDUS

AQUILA

HOROLOGIUM

TUCANA

SAGITTARIUS

CAELUM

RETICULUM

SCUTUM

COLUMBA

DORADO

HYDRUS

PAVO

CORONA
AUSTRALIS

PICTOR

MENSA

TELESCOPIUM

SERPENS
CAUDA

ORION

OCTANS

LEPUS

*South
Celestial Pole*

APUS

ARA

SCORPIUS

CANIS
MAJOR

CARINA

VOLANS

CHAMAELEON

TRIANGULUM
AUSTRALE

OPHIUCHUS

MUSCA

CIRCINUS

NORMA

PUPPIS

CRUX

LUPUS

VELA

CENTAURUS

MONOCEROS

PYXIS

ANTLIA

CANIS MINOR

SERPENS CAPUT

LIBRA

HYDRA

SEXTANS

CRATER

CORVUS

VIRGO

EQUATOR

LEO

ECLIPTIC

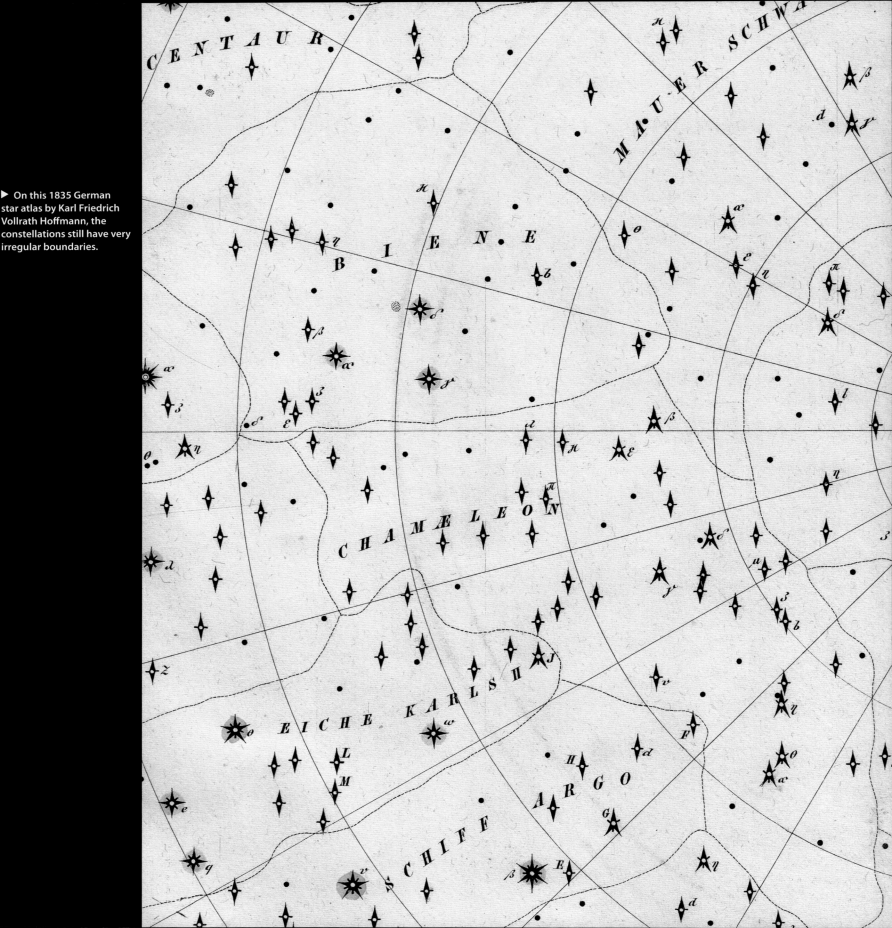

▶ On this 1835 German star atlas by Karl Friedrich Vollrath Hoffmann, the constellations still have very irregular boundaries.

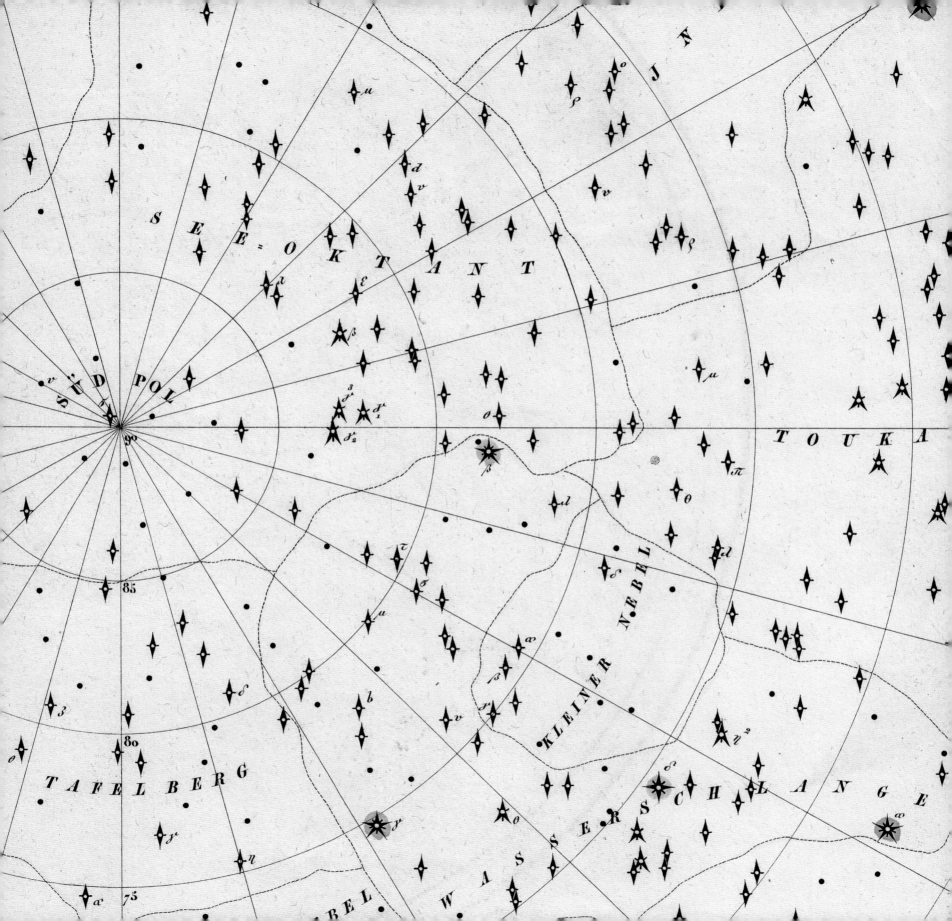

ANDROMEDA

PASSPORT

Latin name: Andromeda

English name: Andromeda

Genitive: Andromedae

Abbreviation: And

Origin: Ptolemy

Area: 722.3 square degrees

Number of naked-eye stars: 152

Bordering constellations: Cassiopeia, Lacerta, Pegasus, Pisces, Triangulum, Perseus

Best visibility: September–October, north of 35° south

LACERTA

PERSEUS

CASSIOPEIA

+50°

+50°

23ʰ

0ʰ

2ʰ

φ

λ

1ʰ

891

Almach
γ

Ross 248

7662

o

+40°

υ Titawin

ν

M110

M31
Andromeda Galaxy

M32

+40°

752

R

μ

23ʰ

2ʰ

▼ A small group of dots near the mouth of the fish in this image from al-Sufi's *Book of Fixed Stars* (964 AD) is actually the Andromeda galaxy.

Mirach β

ANDROMEDA

PEGASUS

TRIANGULUM

PISCES

1ʰ

0ʰ

π

δ

α

+30°

Alpheratz

ANDROMEDA IS A LARGE CONSTELLATION in the northern sky. It depicts the beautiful daughter of King Cepheus and Queen Cassiopeia, the rulers of Ethiopia in Greek mythology. After Cassiopeia had insulted the gods, Cepheus was ordered to sacrifice his daughter to the heinous sea monster Cetus. Andromeda was chained to a sea cliff but was rescued by the half god Perseus. All the protagonists in the story can be found in the night sky.

The constellation is easy to find. First, try to locate the Great Square of Pegasus, which is a conspicuous group of four stars in the (northern) autumn sky. Andromeda stretches toward the left from the upper left corner of the Square (in fact, this star belongs to Andromeda, not to Pegasus).

Andromeda's claim to fame is being host to the Andromeda galaxy, the nearest large spiral galaxy to our own Milky Way, at a distance of a "mere" 2.5 million light-years. On a clear, moonless night, the Andromeda galaxy can just be glimpsed by the unaided eye. The galaxy, also known as M31, is accompanied by two smaller elliptical galaxies, M32 and M110.

▶ Dust clouds from which new stars are born appear pink in this infrared photo of the Andromeda galaxy, obtained by the Spitzer Space Telescope.

▶ Stellar velocities in the core of M32, a small companion of the Andromeda galaxy, reveal the existence of a supermassive black hole in its core, weighing in at a few million times the mass of the sun.

▼ Artist's impression of exoplanet Upsilon Andromedae d, also known as Majriti. A closer-in planet, Saffar, can be seen to the upper left of the star.

▶ In the lower left of this Hubble photo, almost lost between millions of other stars, is the Cepheid variable that Edwin Hubble found in the Andromeda galaxy.

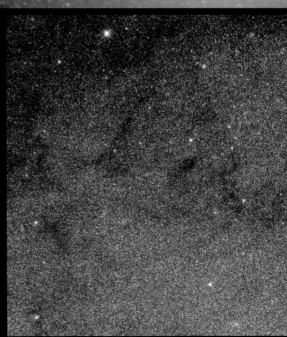

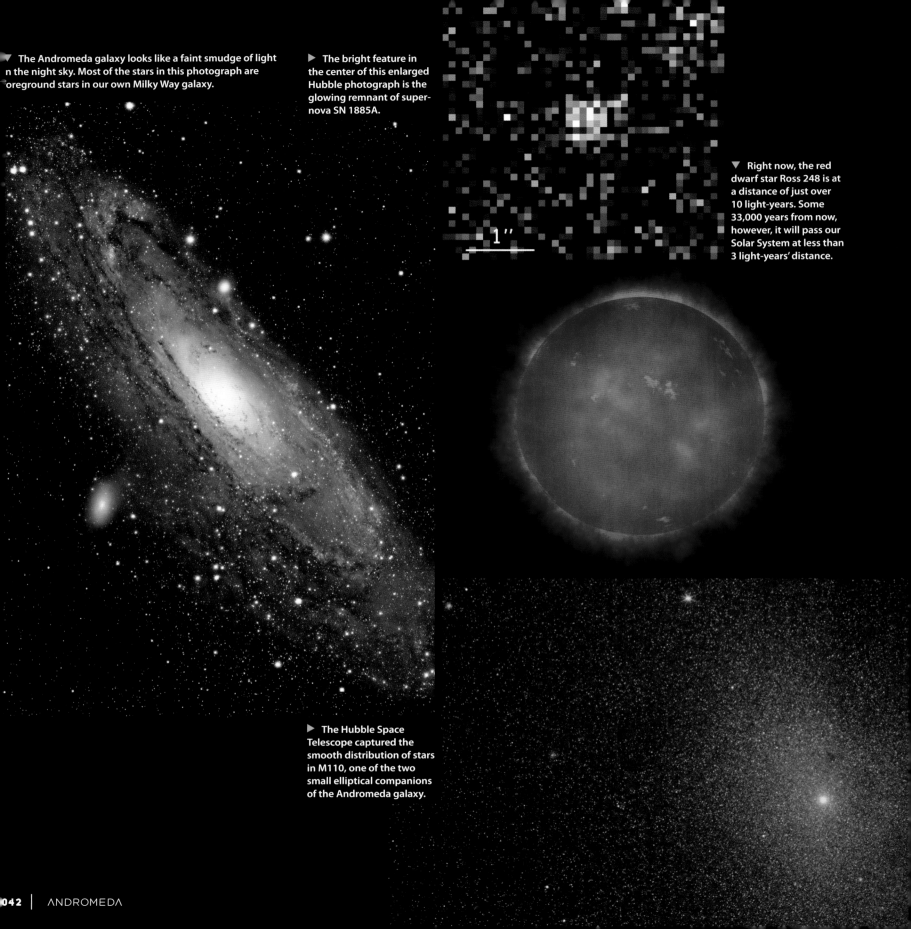

▼ The Andromeda galaxy looks like a faint smudge of light in the night sky. Most of the stars in this photograph are foreground stars in our own Milky Way galaxy.

▶ The bright feature in the center of this enlarged Hubble photograph is the glowing remnant of supernova SN 1885A.

1″

▼ Right now, the red dwarf star Ross 248 is at a distance of just over 10 light-years. Some 33,000 years from now, however, it will pass our Solar System at less than 3 light-years' distance.

▶ The Hubble Space Telescope captured the smooth distribution of stars in M110, one of the two small elliptical companions of the Andromeda galaxy.

964   Persian astronomer Abd al-Rahman al-Sufi publishes his *Book of Fixed Stars*, which contains the first mention of a faint nebula in this constellation—the Andromeda galaxy.

1749   Guillaume Le Gentil (France) discovers M32, a dwarf elliptical galaxy companion to the Andromeda galaxy.

1773   French astronomer Charles Messier is the first to observe the Andromeda galaxy's second companion, M110.

1778   German astronomer Johann Tobias Mayer discovers the binary nature of the star Almach (Gamma Andromedae or γ And). A small telescope reveals a bright, yellow star, accompanied by a dimmer blue one.

1885   Throughout Europe, astronomers notice a "new star" (*stella nova*) in what was then known as the Andromeda Nebula, barely visible to the naked eye. We now know this was actually a supernova—the catastrophic explosion of a dying star. Known as SN 1885A or S And, it was the first supernova seen outside our own Milky Way galaxy, and so far the only one observed in M31.

1923   American astronomer Edwin Hubble discovers a so-called Cepheid variable star in the Andromeda Nebula. This allows him to estimate its distance and to prove once and for all that spiral nebulae are full-fledged galaxies, located well beyond our own Milky Way.

1926   Frank Elmer Ross (USA) measures the large proper motion of a tiny red dwarf star in the Andromeda galaxy, now known as Ross 248. The star is heading toward our sun, which it will pass at a mere 3 light-years, some 33,000 years from now.

1961   British cosmologist and science fiction author Fred Hoyle, together with John Elliot, writes the seven-part BBC TV series *A for Andromeda*, in which an alien radio message from M31 helps scientists to create artificial life on Earth.

1984   The first supermassive black hole is discovered in Andromeda, by American astronomer John Tonry. It resides in the core of the dwarf elliptical galaxy M32.

1996   Geoff Marcy and Paul Butler (USA) discover a Jupiter-like planet orbiting the star Titawin (Upsilon Andromedae or υ And). This is only the second discovery of an exoplanet around a solar-type star. Later, three more planets are found in the system. The innermost three of the four planets are now officially named Saffar, Samh, and Majriti.

2012   Precise measurements of stellar motions in M31, carried out by the Hubble Space Telescope, reveal that the Andromeda galaxy will collide with our Milky Way some 4 billion years from now.

▶ **Both our own Milky Way and M31 will be hugely distorted by tidal forces when the two galaxies crash into each other some 4 billion years from now.**

ANTLIA

PASSPORT

Latin name: Antlia	**Area:** 238.9 square degrees
English name: Air Pump	**Number of naked-eye stars:** 42
Genitive: Antliae	**Bordering constellations:** Hydra, Pyxis, Vela, Centaurus
Abbreviation: Ant	**Best visibility:** April–June, south of 45° north
Origin: de Lacaille	

ANTLIA (THE AIR PUMP) is an inconspicuous constellation south of Hydra (the Sea Serpent). Its brightest star, Alpha Antliae or α Ant, is only 4th magnitude; all other stars in this constellation are fainter still.

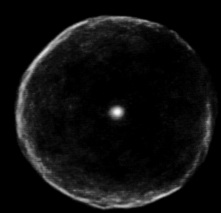

▶ Around 700 BC, the ruddish carbon star U Antliae experienced a brief period of rapid mass loss, resulting in this expanding shell imaged by the ALMA observatory.

TIMELINE

1763 Antlia is introduced as a new constellation by French astronomer Nicolas Louis de Lacaille in his southern star catalog *Coelum Australe Stelliferum*.

1985 Astronomers discover the Antlia dwarf galaxy, a loose collection of stars at a distance of 4.3 million light-years.

2017 Using the Atacama Large Millimeter/submillimeter Array (ALMA) in Chile, astronomers discover a dusty bubble around the old carbon star U Antliae. The expanding shell was formed some 2,700 years ago.

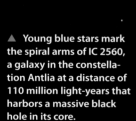

▲ Young blue stars mark the spiral arms of IC 2560, a galaxy in the constellation Antlia at a distance of 110 million light-years that harbors a massive black hole in its core.

▶ The Antlia dwarf galaxy is little more than an overdensity of stars, just beyond the Local Group of galaxies that contains our own Milky Way.

APUS

PASSPORT

Latin name: Apus	**Area:** 206.3 square degrees
English name: Bird of Paradise	**Number of naked-eye stars:** 39
Genitive: Apodis	**Bordering constellations:** Triangulum Australe, Circinus, Musca,
Abbreviation: Aps	Chamaeleon, Octans, Pavo, Ara
Origin: Keyser and de Houtman	**Best visibility:** April–June, south of 5° north

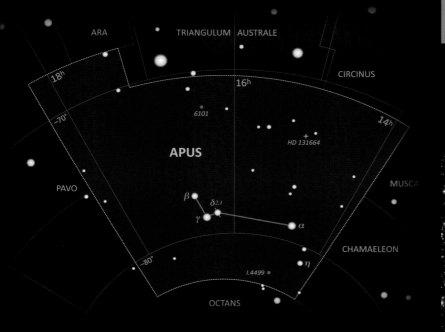

APUS (THE BIRD OF PARADISE) is located close to the south celestial pole. As a result, it is visible only from the southern hemisphere. But even then, it's hard to find: It contains almost no bright stars.

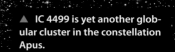

▲ **IC 4499 is yet another globular cluster in the constellation Apus.**

▲ **Apus — the Bird of Paradise — shows off its beautiful feathers in the 1603 star atlas** *Uranometria* **by Johann Bayer.**

◄ **Tens of thousands of stars populate the core of the globular cluster NGC 6101.**

TIMELINE

1598   Flemish astronomer Petrus Plancius first depicts the constellation on a celestial globe, based on descriptions by Dutch sailors Pieter Dirkszoon Keyser and Frederick de Houtman.

1826   From Australia, Scottish astronomer James Dunlop discovers NGC 6101, a globular cluster at a distance of some 50,000 light-years.

2008   European astronomers discover a brown dwarf companion to the star HD 131664—a "failed star" about as large as the giant planet Jupiter, but over twenty times as massive.

AQUARIUS

PASSPORT

Latin name: Aquarius	**Area:** 979.9 square degrees
English name: Water Bearer	**Number of naked-eye stars:** 172
Genitive: Aquarii	**Bordering constellations:** Pegasus, Equuleus, Delphinus, Aquila, Capricornus, Piscis Austrinis, Sculptor, Cetus, Pisces
Abbreviation: Aqr	
Origin: Ptolemy	**Best visibility:** July–September, between 65° north and 85° south

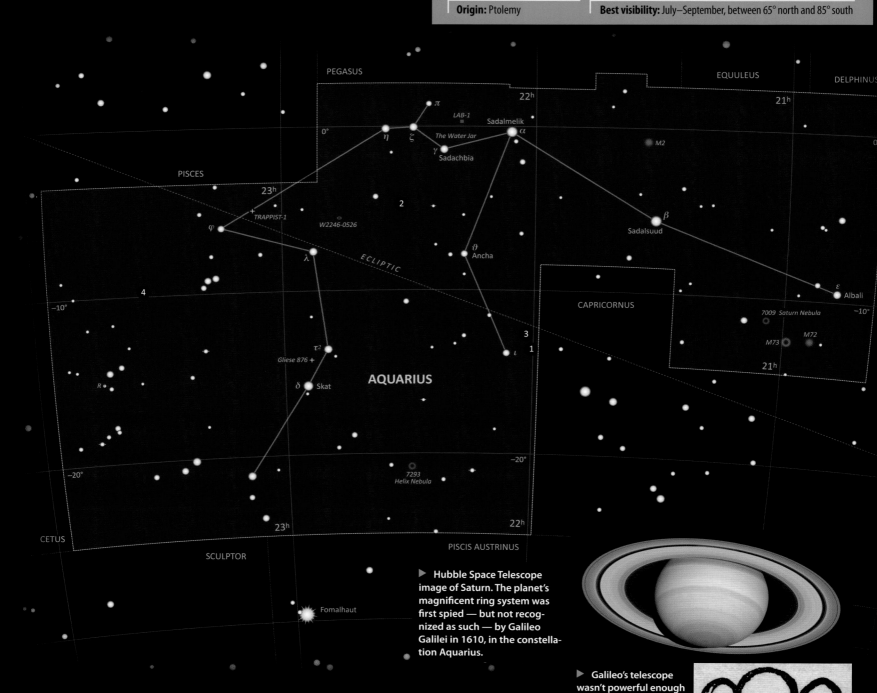

▶ Hubble Space Telescope image of Saturn. The planet's magnificent ring system was first spied — but not recognized as such — by Galileo Galilei in 1610, in the constellation Aquarius.

▶ Galileo's telescope wasn't powerful enough to reveal the true nature of Saturn's rings.

AQUARIUS (THE WATER BEARER) is one of the largest constellations in the sky. It's also one of the oldest, dating back to the early Sumerians, for whom it represented Enki, the god of water. Although it has a quite conspicuous shape, the constellation may be hard to recognize from populated areas because it contains just a handful of really bright stars.

Aquarius is best visible between August and November. It occupies the triangular area between the Great Square of Pegasus, the bright star Altair in the constellation Aquila (the Eagle), and the equally bright star Fomalhaut in the constellation Piscis Austrinus (the Southern Fish).

Aquarius is one of the twelve constellations of the zodiac. The sun crosses the constellation each year between February 16 and March 12, and the moon and the naked-eye planets regularly sit within its borders. Little wonder, then, that many solar system observations and discoveries have been made in this constellation.

◀ **The giant planet orbiting the dwarf star Gliese 876 in Aquarius is accompanied by several moons in this artist's rendition.**

▶ **Thin ribbons of cold gas show up in this infrared view of the Helix Nebula, captured by the European VISTA telescope.**

▶ **The extremely luminous galaxy W2246-0526 produces so much radiation that it is almost blowing itself apart.**

1610 ❶ Galileo Galilei points his homemade telescope at the planet Saturn, close to the star Iota Aquarii, or ι Aqr, and discovers that it has two apparent companions on either side of the planet. Only later would it turn out that these are actually the tips of the planet's ring system. Due to the poor optical quality of his telescope, Galileo didn't recognize the rings for what they are.

1804 ❷ Juno, the third asteroid to be found (after Ceres and Pallas), is discovered by Karl Harding in Germany while it is slowly traversing the constellation Aquarius.

1824 In Göttingen, Germany, Karl Harding discovers the Helix Nebula (NGC 7293) in Aquarius, one of the nearest planetary nebulae in our Milky Way galaxy.

1846 ❸ Based on calculations by French mathematician Urbain Le Verrier, Johann Gottfried Galle of the Berlin Observatory in Germany discovers the planet Neptune, in the same part of the sky where Galileo observed Saturn 236 years earlier. Le Verrier derived the planet's position from its gravitational influence on the motion of Uranus. As a result, Neptune has been described as the "writing desk planet."

1877 ❹ In the eastern part of the constellation, American astronomer Asaph Hall discovers the two tiny moons of Mars, Phobos and Deimos. The largest crater on Phobos, first imaged by the Mariner 9 spacecraft in 1972, is named Stickney. The name honors Hall's wife, Angeline Stickney, who had encouraged him not to give up the search for Martian moons.

1998 A planet is found orbiting the star Gliese 876 in Aquarius. It's the first time astronomers discover a planet in orbit around a red dwarf star. Today we know that many red dwarf stars are accompanied by relatively small planets.

2000 American astronomers discover a huge cloud of hydrogen gas in the constellation Aquarius, at a distance of 11.5 billion light-years. The cloud, catalogued as Lyman-alpha blob 1, or LAB-1, is three times the size of our Milky Way galaxy. It is probably a primordial mass of hydrogen from which new galaxies will condense.

2015 Using the small TRAPPIST telescope in Chile, Belgian astronomers discover three planets orbiting the red dwarf star TRAPPIST-1 in Aquarius, which is only 40 light-years away. In 2017, four more planets are found. All seven planets are comparable in size to Earth; three of them may be habitable.

2016 In the northeastern part of the constellation, the ALMA Observatory (Atacama Large Millimeter/submillimeter Array) in northern Chile discovers the most luminous galaxy known in the universe. Known as W2246-0526, it is a so-called Hot DOG—a hot dust-obscured galaxy. The extremely remote galaxy produces as much light as 350 trillion suns. In its core is a supermassive black hole, which powers the turbulent behavior of the galaxy's interstellar gas.

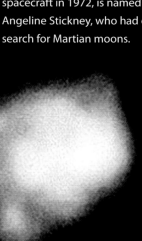

◀ This is the best photograph available of the irregularly shaped asteroid Juno, which is about 235 kilometers across.

▶ The planet Neptune was discovered by Johann Gottfried Galle (left position) close to where Urbain Le Verrier had predicted it to be (right position).

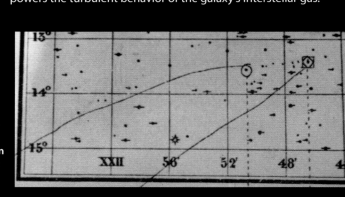

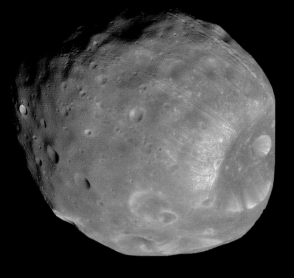

▼ The large crater on the lower right in this photo of the Martian moon Phobos is called Stickney, after the wife of Asaph Hall, who discovered Phobos and Deimos in 1877.

▲ Seven Earth-sized planets orbit the tiny red dwarf star TRAPPIST-1, which is just 40 light-years away in the constellation Aquarius.

◀ Almost at the edge of the observable universe is this giant blob of hydrogen gas, known as Lyman-alpha blob 1, or LAB-1 for short.

AQUILA

PASSPORT

Latin name: Aquila

English name: Eagle

Genitive: Aquilae

Abbreviation: Aql

Origin: Ptolemy

Area: 652.5 square degrees

Number of naked-eye stars: 124

Bordering constellations: Sagitta, Hercules, Ophiuchus, Serpens Cauda, Scutum, Sagittarius, Capricornus, Aquarius, Delphinus

Best visibility: July–September, between 75° north and 70° south

▶ The orbit of the Hulse-Taylor binary is slowly shrinking as gravitational waves carry away energy.

VULPECULA

SAGITTA

19ʰ

20ʰ

PSR B1913+ 16

Hen 2-428

ε

Okab

ζ

HERCULES

OPHIUCHUS

π

γ   Tarazed

6709

+10°

+10°

α   Altair

R

β   Alshain

SERPENS CAUDA

DELPHINUS

δ

η

Nova 1918

0°

0°

ϑ

Smith's Cloud (Center)

AQUILA

λ

SCUTUM

CAPRICORNUS   −10°

−10°

20ʰ

19ʰ

SAGITTARIUS

AQUILA (THE EAGLE) is a prominent constellation in the northern summer sky. Its brightest star, Altair, forms the southernmost tip of the giant Summer Triangle, whose other vertices are Deneb in Cygnus (the Swan) and Vega in Lyra (the Lyre).

The constellation was already listed by Greek astronomer Claudius Ptolemy. It represents the eagle carrying the thunderbolts of Zeus and features in the story of Ganymede, a Trojan prince who was brought to Mount Olympus to serve the gods.

Looking toward Aquila, one also looks into the central plane of our Milky Way galaxy. On a moonless night, the Milky Way in Aquila can be seen as a faint band of soft light, dissected by a conspicuous rift of dark dust.

▶ Stretched out by tidal forces, Smith's Cloud is a giant mass of hydrogen approaching our Milky Way galaxy.

► Nova Aquilae 1918, as imaged at the Ukrainian National Observatory.

Nova Aquilae 18 июня 1918 года

▲ Artistic rendition of the rapidly rotating and strongly flattened star Altair.

▼ At the heart of the planetary nebula Henize 2-428 is a binary white dwarf.

TIMELINE

1918 In June 1918, a bright nova (exploding star) appears in Aquila. Known as Nova Aquilae 1918, it briefly becomes the third-brightest star in the night sky.

1963 Using the 82-foot radio telescope in Dwingeloo, the Netherlands, Gail Bieger-Smith discovers a giant, fast-moving cloud of hydrogen gas in Aquila, with a total space velocity of about 185 miles per second. Some 30 million years from now, Smith's Cloud will collide and merge with the Milky Way galaxy.

1974 American astronomers Russell Hulse and Joe Taylor discover a pulsar—a very compact and rapidly rotating star—in a binary system: PSR B1913+16. Later observations of the Hulse-Taylor binary indirectly prove the existence of gravitational waves— Einstein's famous ripples in space-time.

2007 By linking up four telescopes, American astronomers succeed in actually imaging the surface of Altair, which is at a distance of just 17 light-years. The star is strongly flattened as a result of its very fast rotation.

2014 Thanks to the resolving power of the European Southern Observatory's Very Large Telescope, astronomers discover that the central star of the planetary nebula Henize 2-428 (or Hen 2-428 for short) is actually a binary white dwarf. Hundreds of millions of years from now, the two stars will merge and explode as a luminous supernova.

ARA

PASSPORT

Latin name: Ara		**Area:** 237.1 square degrees	
English name: Altar		**Number of naked-eye stars:** 71	
Genitive: Arae		**Bordering constellations:** Scorpius, Norma, Triangulum Australe, Apus, Pavo, Telescopium, Corona Australis	
Abbreviation: Ara			
Origin: Ptolemy		**Best visibility:** May–July, south of 20° north	

ARA (THE ALTAR) can be found south of the tail of Scorpius (the Scorpion). Its stars are relatively bright, but it lacks a very conspicuous shape, making it hard to recognize. Its name refers to the fact that the Milky Way seems to rise up from the constellation like smoke from an altar.

TIMELINE

1961 Swedish astronomer Bengt Westerlund discovers a large, young cluster of stars in Ara at a distance of 11,500 light-years, now known as Westerlund 1. One of the stars in the cluster appears to be some two thousand times as large as our sun.

1994 The Hubble Space Telescope succeeds in imaging the Stingray Nebula (Hen 3-1357). It is the youngest known planetary nebula; in 1971, it wasn't yet hot enough to glow by itself.

2004 The first super-Earth (or mini-Neptune) is found, orbiting the star Mu Arae, or μ Ara, as the innermost planet of the star's four-planet system.

▲ The Stingray Nebula is the youngest planetary nebula known.

◀ New stars are born in RCW 108, a dusty stellar nursery in Ara.

▶ Star cluster Westerlund 1 contains one of the largest stars known.

ARIES

PASSPORT

Latin name: Aries	**Area:** 441.4 square degrees
English name: Ram	**Number of naked-eye stars:** 86
Genitive: Arietis	**Bordering constellations:** Perseus, Triangulum, Pisces, Cetus, Taurus
Abbreviation: Ari	
Origin: Ptolemy	**Best visibility:** September–November, north of 55° south

ARIES (THE RAM) is the first of the twelve constellations of the zodiac. It is easy to spot, west of Taurus (the Bull). Its brightest star is called Hamal, from the Arab *Ras al-hamal* (the head of the ram).

◄ The constellation Aries (the Ram), as it appears in an eleventh-century manuscript.

◄ The strongly disturbed galaxy NGC, imaged by the Liverpool Telescope on La Palma, Canary Islands.

TIMELINE

1977   ❶ American astronomer Charles Kowal discovers the first centaur, Chiron, at the southern edge of Aries. Centaurs are icy asteroids orbiting within the realm of the outer planets, well beyond the asteroid belt. Chiron appears to be surrounded by two sharply defined rings.

2003   Within one year, two supernova explosions are observed in the galaxy NGC 772. The galaxy has a peculiar spiral arm, probably as a result of tidal forces from a companion dwarf galaxy.

▼ Chiron, the first centaur, may be surrounded by two rings of icy particles.

AURIGA

PASSPORT

Latin name: Auriga	**Area:** 657.4 square degrees
English name: Charioteer	**Number of naked-eye stars:** 152
Genitive: Aurigae	**Bordering constellations:** Camelopardalis, Perseus, Taurus, Gemini, Lynx
Abbreviation: Aur	
Origin: Ptolemy	**Best visibility:** November–January, north of 30° south

AURIGA (THE CHARIOTEER) is a very prominent constellation in the northern winter sky. Its brightest star, Capella, is the sixth-brightest star in the night sky, and is part of the so-called Winter Hexagon.

To the Greeks, Auriga represented the mythological hero Erichthonius, who supposedly invented the four-horse chariot. He is often depicted with a goat and two baby goats; Capella is sometimes called the Goat Star. The easternmost part of Auriga was once a separate constellation, Telescopium Herschelii (Herschel's Telescope), to honor Herschel's discovery of the planet Uranus in 1781.

As Auriga lies in the plane of the Milky Way, it contains a large number of open star clusters, some of which are visible in binoculars.

▶ Once every 27 years, the light of the giant star Epsilon Aurigae is partly obscured by a thick disk of dust surrounding a smaller companion star.

▼ M37 is one of the brightest open star clusters in Auriga.

TIMELINE

1891   Thomas Anderson, an amateur astronomer from Scotland, discovers a nova outburst in Auriga. For a number of months, Nova Auriga 1891 was easily visible to the naked eye.

1899   American astronomer William Campbell discovers that Capella is a binary star, consisting of two yellow giants orbiting each other once every 104 days at a distance of 111 million kilometers. At a much larger distance, the pair is now known to be orbited by yet another binary, making Capella a quadruple system.

1904   On the basis of his own observations, German astronomer Hans Ludendorff concludes that the star Almaaz (Epsilon Aurigae or ε Aur) is an eclipsing binary, where one star periodically obscures the other. This only happens once every 27 years. Astronomers now believe that the companion star is surrounded by a huge, opaque disk of dust, which explains the erratic variability of the primary.

2012   In the southern part of Auriga, the first (and so far only) repeating fast radio burst is discovered in observations with the Very Large Array radio telescope in New Mexico. While the nature of fast radio bursts is still a mystery, FRB 121102 has been traced back to the outer parts of a faint dwarf galaxy more than 3 billion light-years away. The extremely brief but energetic bursts probably occur on or near a strongly magnetized neutron star — the remnant of a supernova explosion.

2013   The massive galaxy cluster MACSJ0717.5+3745, in the eastern part of Auriga, is first observed as part of the Hubble Space Telescope's Frontier Fields program. At a distance of 5.4 billion light-years, MACSJ0717 is a smash-up of four smaller galaxy clusters.

▲ Hubble Frontier Fields observation of galaxy cluster MACSJ0717. The images of background galaxies are distorted by the gravity of the cluster.

▲ In 1627, German astronomer Julius Schiller depicted Auriga as Saint Jerome (Hieronymus) in his *Coelum Stellatum Christianum*.

▼ Strong polarization of radio waves shed light on the magnetic environment of the repeating fast radio burst FRB 121102.

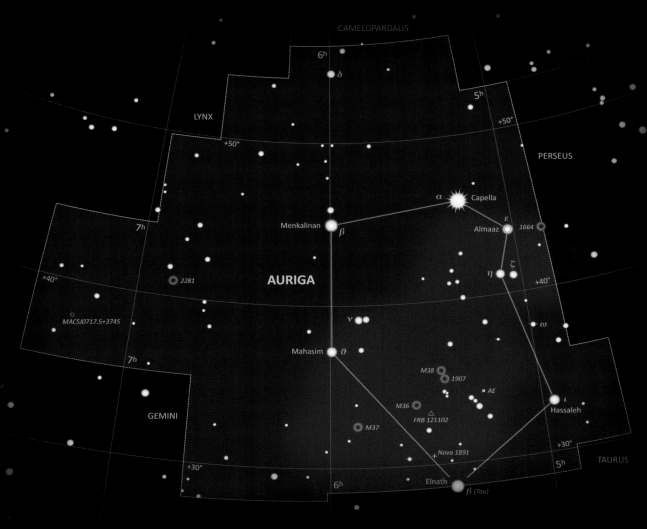

BOÖTES

PASSPORT

Latin name: Boötes	**Area:** 906.8 square degrees
English name: Herdsman	**Number of naked-eye stars:** 144
Genitive: Boötis	**Bordering constellations:** Draco, Ursa Major, Canes Venatici, Coma Berenices, Virgo, Serpens Caput, Corona Borealis, Hercules
Abbreviation: Boo	
Origin: Ptolemy	**Best visibility:** April–May, north of 35° south

BOÖTES (THE HERDSMAN) is a large constellation in the northern sky, in the shape of a huge kite. At the bottom tip of the kite is Arcturus, the brightest star in the northern hemisphere and the fourth-brightest star in the entire night sky.

Boötes is sometimes described as a plowman—to the Greeks, the nearby Big Dipper was also known as the Plow. Extending the curved handle of the Big Dipper takes you to Arcturus, and then on to Spica in the constellation Virgo (the Virgin).

Part of Boötes was once regarded as a separate constellation, Quadrans Muralis (the Mural Quadrant)—a designation that still survives in the name of the early January meteor shower the Quadrantids.

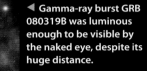

◀ **Gamma-ray burst GRB 080319B was luminous enough to be visible by the naked eye, despite its huge distance.**

TIMELINE

1718 By comparing the position of Arcturus (and a few other stars) with older observations, English astronomer Edmund Halley discovers the proper motion of stars across the sky—evidence that stars are moving through space. The velocity of Arcturus turns out to be 75 miles per second.

1981 At a distance of 700 million light-years, the Boötes Void is discovered— a huge region in intergalactic space, some 300 million light-years across, that is almost devoid of galaxies. It is the first cosmic void that is recognized as such.

1996 American astronomers Geoff Marcy and Paul Butler discover a hot giant planet orbiting the star Tau Boötis, or τ Boo—one of the first extrasolar planets found, at a distance of just 51 light-years from Earth.

2008 On March 19, the most luminous gamma-ray burst ever, GRB 080319B, is detected by NASA's Swift satellite in the constellation Boötes. Despite its distance of some 7.5 billion light-years, the titanic stellar explosion is briefly visible to the naked eye.

2017 The Wide-field Infrared Survey Explorer discovers the most distant quasar found to date. Known as ULAS J1342+0928 and located in the extreme southwest corner of Boötes, the remote galaxy harbors a supermassive black hole weighing in at 800 million times the mass of the sun.

▲ **Arcturus is the fourth brightest star in the night sky.**

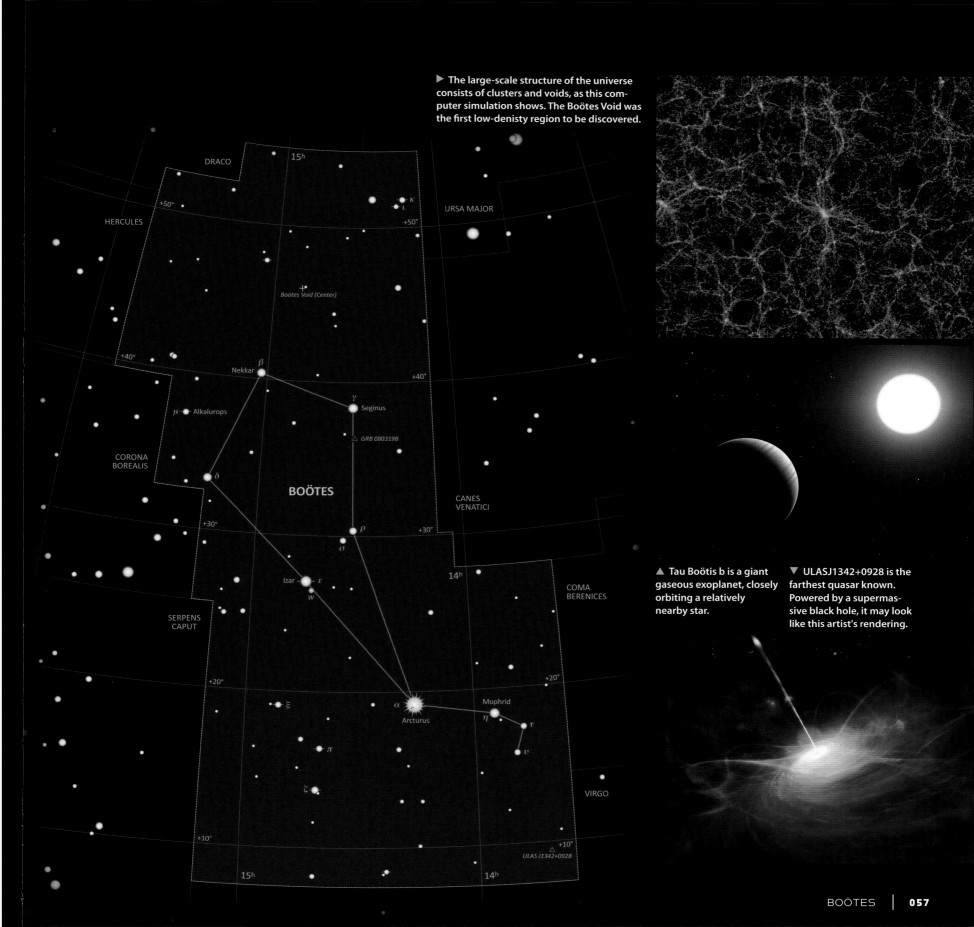

▶ The large-scale structure of the universe consists of clusters and voids, as this computer simulation shows. The Boötes Void was the first low-denisty region to be discovered.

DRACO

15ʰ

+50°

HERCULES

+50°

URSA MAJOR

+50°

κ
ι

+40°

Boötes Void (Center)

+40°

β
Nekkar

+40°

γ
Seginus

μ — Alkalurops

△ GRB 080319B

CORONA
BOREALIS

δ

BOÖTES

CANES
VENATICI

+30°

ρ

+30°

σ

14ʰ

Izar — ε

COMA
BERENICES

w

SERPENS
CAPUT

+20°

+20°

ο
α
Arcturus

Muphrid

η
τ

π

υ

ζ

VIRGO

+10°

+10°
△
ULAS J1342+0928

15ʰ

14ʰ

▲ Tau Boötis b is a giant gaseous exoplanet, closely orbiting a relatively nearby star.

▼ ULASJ1342+0928 is the farthest quasar known. Powered by a supermassive black hole, it may look like this artist's rendering.

CAELUM

PASSPORT

Latin name: Caelum	Area: 124.9 square degrees
English name: Chisel	Number of naked-eye stars: 20
Genitive: Caeli	Bordering constellations: Eridanus, Horologium, Dorado, Pictor, Columba, Lepus
Abbreviation: Cae	
Origin: de Lacaille	Best visibility: November–December, south of 40° north

CAELUM (THE CHISEL) is one of the smallest and most inconspicuous constellations in the sky. Even its brightest star, Alpha Caeli, or α Cae, is visible to the naked eye only under favorable circumstances.

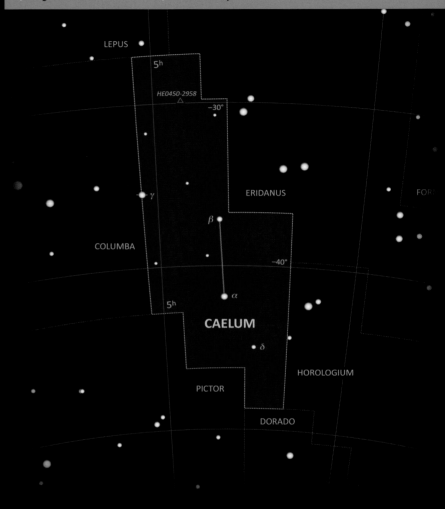

▲ At a distance of 66 light-years, Alpha Caeli is the brightest star in the small constellation.

▼ The relatively nearby quasar HE0450-2958 in Caelum appears to be an active supermassive black hole without a host galaxy.

TIMELINE

1763 Nicolas Louis de Lacaille introduces Caelum as a new southern sky constellation, representing an engraver's chisel.

2005 French astronomers discover a relatively close quasar in Caelum, HE0450-2958, at a distance of approximately 3 billion light-years. Quasars are basically supermassive black holes that spew jets of matter into space. Remarkably, HE0450-2958 appears to lack a host galaxy.

▼ Caelum appears as Caela Scalptoris in Johann Bode's *Uranographia* (1801).

CAMELOPARDALIS

PASSPORT

Latin name: Camelopardalis	**Area:** 756.8 square degrees
English name: Giraffe	**Number of naked-eye stars:** 152
Genitive: Camelopardalis	**Bordering constellations:** Cepheus, Cassiopeia, Perseus, Auriga, Lynx, Ursa Major, Draco, Ursa Minor
Abbreviation: Cam	
Origin: Plancius	**Best visibility:** November–December, north of the equator

▲ Camelopardalis, as it appears in *Urania's Mirror*, an 1825 set of astronomical cards, together with two now obsolete constellations.

CAMELOPARDALIS (THE GIRAFFE) is a large but remarkably inconspicuous constellation in the northern sky, more or less between the Pole Star and Capella. It's quite a challenge to actually find it, since it doesn't contain any bright stars at all.

◄ The extremely faint speck of light on this Hubble photo of a galaxy cluster in Camelopardalis is MACS0647-JD, one of the farthest galaxies known.

TIMELINE

1612 — Flemish astronomer and cartographer Petrus Plancius introduces Camelopardalis as a new constellation on a celestial globe.

1788 — William Herschel discovers the spiral galaxy NGC 2403, at some 8 million light-years' distance in Camelopardalis. In 2004, supernova SN 2004DJ explodes in this galaxy.

2012 — Thanks to gravitational lensing by a foreground galaxy cluster, astronomers discover the galaxy MACS0647-JD, which is one of the most distant galaxies known, probably at some 13.3 billion light-years away. The baby galaxy is just 600 light-years across and contains approximately a billion stars.

▲ Supernova 2004DJ (top) lights up in NGC 2403, one of the many spiral galaxies in Camelopardalis.

CANCER

PASSPORT

Latin name: Cancer	**Area:** 505.9 square degrees
English name: Crab	**Number of naked-eye stars:** 104
Genitive: Cancri	**Bordering constellations:** Lynx, Gemini, Canis Minor, Hydra, Leo, Leo Minor (corner)
Abbreviation: Cnc	
Origin: Ptolemy	**Best visibility:** January–February, north of 55° south

▲ Jupiter's moons Io (left) and Europa float above the planet's turbulent cloud tops in this Voyager 1 image.

▶ Voyager 2 imaged Jupiter's Great Red Spot in July 1979, when the giant planet was in Cancer, as seen from Earth.

▼ At the center of this image, shot by Voyager 1 in March 1979, is the doughnut-shaped volcano Prometheus. Voyager 1 was the first spacecraft to detect active volcanism beyond Earth.

CANCER (THE CRAB) is one of the smallest and least conspicuous constellations of the zodiac. The sun moves through this constellation between July 20 and August 10. On dark northern winter and spring nights, Cancer can be found midway between Gemini (the Twins) and Leo (the Lion). Even though its main stars are relatively faint, it has a very recognizable shape.

The best binocular sight in Cancer is the Beehive open star cluster, also known as Praesepe (the Crib) or M44, which is just visible with the naked eye, under favorable circumstances. Since Cancer is part of the zodiac, photogenic conjunctions of planets with the star cluster are quite common.

▲ Amateur astronomer Bob Franke captured this colorful image of the open star cluster M44, also known as Praesepe (the Crib) or the Beehive Cluster.

▶ Planet 55 Cancri f (foreground), discovered in 2007, may have moons that are supportive of life.

TIMELINE

1609 Galileo Galilei is the first one to observe individual stars in the Beehive Cluster, with his homemade telescope. Earlier astronomers had listed Praesepe as a nebulous cloud.

1779 German astronomer Johann Köhler discovers another open star cluster in Cancer, now known as M67. The cluster has been found to contain a relatively large number of so-called hot Jupiters—giant planets in tight orbits.

1979 ❶❷ NASA's Deep Space Network is aimed at Cancer to receive signals from the Voyager 1 and Voyager 2 spacecraft during their encounters with Jupiter and its moons, which occur in this constellation on March 5 and July 9, respectively.

1996 The first of at least five planets is discovered in orbit around the star 55 Cancri, or ρ1 Cnc, which is 41 light-years away, in the northern part of the constellation.

2002 In a highly criticized experiment, astronomers Ed Fomalont and Sergei Kopeikin try to determine the speed of gravity by measuring the bending of light from the remote quasar QSO J0842+1835 by the gravity of Jupiter, which is again moving through Cancer.

2003 A radio message known as Cosmic Call 2 is sent to the planetary system of the star 55 Cancri, using the 229-foot radar dish in Yevpatoria (Crimea). It is due to arrive in 2044.

CANES VENATICI

PASSPORT

Latin name: Canes Venatici	**Area:** 465.2 square degrees
English name: Hunting Dogs	**Number of naked-eye stars:** 59
Genitive: Canum Venaticorum	**Bordering constellations:** Ursa Major, Coma Berenices, Boötes
Abbreviation: CVn	**Best visibility:** March–April, north of 35° south
Origin: Hevelius	

CANES VENATICI (THE HUNTING DOGS) is a small but rather conspicuous constellation just south of the handle of the Big Dipper. It represents Asterion and Chara, the two dogs of the herdsman Boötes. However, the dogs were introduced as a separate constellation only in the late seventeenth century.

Canes Venatici is little more than two intermediate-bright stars. But since they're located so close to the famous Big Dipper, they're quite easy to spot. The brightest star, Alpha Canes Venatici, or α CVn, is called Cor Caroli (Heart of Charles), after King Charles I of England.

The constellation is home to the Whirlpool galaxy (M51), one of the most famous and beautiful spiral galaxies in the entire sky, at a distance of 28 million light-years. While the Whirlpool is invisible to the naked eye, a simple pair of binoculars is enough to reveal it as a small, circular smudge of light.

▶ **X-rays and radio waves from M106 are color-coded blue and pink, respectively, revealing the anomalous spiral arms of the galaxy, which is 23 million light-years away.**

TIMELINE

1687 Canes Venatici is introduced as a new constellation by Polish astronomer Johannes Hevelius in his star atlas *Firmamentum Sobiescianum*.

1764 French astronomer Charles Messier discovers another "nebula" in Canes Venatici, which he lists as the third entry (M3) in his *Catalogue des Nébuleuses et des Amas d'Étoiles* (Catalog of Nebulae and Star Clusters). It later turns out to be a giant globular cluster almost 40,000 light-years away, containing about half a million stars.

1781 Pierre Méchain, contemporary, compatriot, and colleague of Charles Messier, discovers the galaxy M106, in the northwestern part of Canes Venatici. It displays warped spiral arms and an extremely energetic and luminous core.

1845 The Irish nobleman William Parsons, the Third Earl of Rosse, is the first to observe spiral structure in a nebula, M51. Lord Rosse likens the appearance of the nebula to a whirlpool.

2005 To celebrate the Hubble Space Telescope's fifteenth anniversary, astronomers use Hubble's Advanced Camera for Surveys to produce a giant 90-megapixel mosaic of the Whirlpool galaxy—the largest-ever Hubble photo at that time.

2009 On April 29, the Swift spacecraft detects the most distant gamma-ray burst ever detected. GRB 090429B is so far away that the radiation we receive on Earth was emitted when the universe was just some 600 million years old—less than 5 percent of its present age.

◀ Lord Rosse's 1845 sketch of spiral structure in what came to be known as the Whirlpool Nebula.

▲ This giant Hubble Space Telescope mosaic image captures the beautiful Whirlpool galaxy (M51) and its smaller companion NGC 5194.

14ʰ

13ʰ

+50°

BOÖTES

M51
Whirlpool Galaxy

M106

URSA MA

4449

+40°

M63

M94

Chara
β

+40°

Cor Caroli
α

△ GRB 090429B

CANES VENATICI

4631

13ʰ

14ʰ

+30°

COMA BERENICES

M3

◀ Half a million distant suns swarm around in the globular cluster M3, at the southernmost edge of Canes Venatici.

▶ Canes Venatici appears (mirror-reversed) as one of seven new constellations in the 1687 star atlas of Johannes Hevelius.

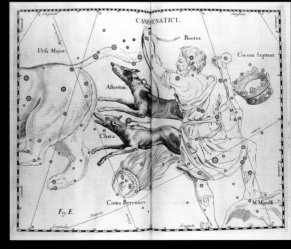

CANIS MAJOR

PASSPORT

Latin name: Canis Major	**Area:** 380.1 square degrees
English name: Greater Dog	**Number of naked-eye stars:** 147
Genitive: Canis Majoris	**Bordering constellations:** Monoceros, Lepus, Columba, Puppis
Abbreviation: CMa	**Best visibility:** December–January, south of 55° north
Origin: Ptolemy	

CANIS MAJOR (THE GREATER DOG) is one of the most prominent constellations in the northern winter sky. It is home to Sirius, the brightest star in the entire night sky. Sirius can be found to the lower left of the famous constellation Orion. Its large apparent brightness is mostly due to its proximity: Sirius is only 8.6 light-years away. Moreover, it is twice as large and as massive as our own sun, and twenty-five times more luminous.

Apart from Sirius, Canis Major contains quite a number of bright stars, making it a very conspicuous constellation, despite its relatively small size. The band of the Milky Way runs through its northeastern part, and a number of star clusters can be found within the constellation's borders, including the open cluster M41.

In Greek mythology, Canis Major was one of the hunting dogs of Orion, although it has also been identified with other mythological dogs. To the Egyptians, the yearly appearance of Sirius (also known as the Dog Star) in the predawn sky marked the imminent flooding of the river Nile.

◀ To the lower left in this Hubble image of the bright star Sirius is its faint companion white dwarf, Sirius B.

▶ A distant spiral galaxy is seen way beyond the small and relatively nearby dwarf galaxy ESO 489-56, in the southwestern part of Canis Major.

▲ The open star cluster M41, a few degrees due south of Sirius, at a distance of 2,300 light-years.

◄ The Canis Major Dwarf galaxy (the small white spot above the center of the image) may be nothing more than an overdensity in a stellar stream (pink) circling around our Milky Way galaxy (blue).

TIMELINE

1698   In his book *Cosmotheoros* (published posthumously), Dutch astronomer Christiaan Huygens describes his method to derive stellar distances by comparing a star's apparent brightness to the brightness of the sun. For Sirius, Huygens arrives at a distance of 0.4 light-years—way too small.

1844   The wobbling motion of Sirius leads German astronomer Friedrich Bessel to conclude that it must have an invisible companion star in a fifty-year orbit.

1862   Testing a newly built 18.5-inch telescope in Cambridgeport, Massachusetts, Alvan Clark actually spots the companion star, which is now known as Sirius B. It is the second white dwarf ever discovered—an extremely compact star, about as massive as our own sun, but hardly larger than the Earth.

1938   French anthropologist Marcel Griaule publishes his study of the West African Dogon people, claiming that they have known about the existence of a companion star to Sirius in a fifty-year orbit for many generations. Later suggestions that the Dogon received this knowledge from visiting Sirian aliens are little more than pseudoscience, though.

2003   Astronomers report their discovery of the Canis Major Dwarf galaxy at a distance of some 25,000 light-years. However, it is unclear whether or not the stellar swarm is a real dwarf galaxy or just a slight overdensity of stars in the outskirts of the Milky Way.

► While Sirius A (left in this artist's impression) is twice the size or our own sun, Sirius B (right) is a compact white dwarf about as small as the Earth.

CANIS MINOR

PASSPORT

Latin name: Canis Minor	**Area:** 183.4 square degrees
English name: Lesser Dog	**Number of naked-eye stars:** 47
Genitive: Canis Minoris	**Bordering constellations:** Gemini, Monoceros, Hydra, Cancer
Abbreviation: CMi	**Best visibility:** December–February, between 85° north and 75° south
Origin: Ptolemy	

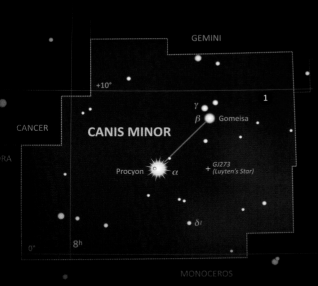

▲ **Luyten's Star (GJ273) in Canis Minor is a red dwarf accompanied by two planets.**

CANIS MINOR (THE LESSER DOG) is a small constellation with just two prominent stars: Procyon —among the ten brightest stars in the night sky —and Gomeisa (Beta Canis Minoris or β CMi). The name *Procyon* means "preceding the dog," which refers to the fact that the star—at least in ancient Greece—rose about an hour earlier than Sirius, the Dog Star, in Canis Major (the Greater Dog).

TIMELINE

1896   At Lick Observatory in California, John Schaeberle discovers Procyon B, the faint white dwarf companion of Procyon.

1935   Dutch astronomer Willem Luyten measures the high proper motion of the red dwarf star GJ273, also known as Luyten's Star. The star is just over a light-year away from Procyon.

1982   ❶ Three years before it becomes visible with the naked eye, Halley's Comet is recovered with one of the first CCD cameras ever used in astronomy, by David Jewitt and Edward Danielson. Every seventy-six years, the comet approaches the inner solar system from the direction of Canis Minor.

2017   Two planets are discovered orbiting Luyten's Star. One, GJ273b, is a super-Earth in the star's habitable zone.

▶ **Recovery image of Halley's Comet (circled), obtained on October 16, 1982, with one of the first electronic cameras ever used in astronomy.**

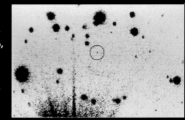

▲ **Canis Minor (the Lesser Dog) sits on the back of Monoceros (the Unicorn) in Alexander Jamieson's 1822 star atlas. Canis Major (the Greater Dog) is at the bottom of the chart.**

CAPRICORNUS

PASSPORT

Latin name: Capricornus	**Area:** 413.9 square degrees
English name: Sea Goat	**Number of naked-eye stars:** 81
Genitive: Capricorni	**Bordering constellations:** Aquarius, Aquila, Sagittarius, Microscopium, Piscis Austrinus
Abbreviation: Cap	
Origin: Ptolemy	**Best visibility:** July–August, south of 60° north

CAPRICORNUS (THE SEA GOAT) is a rather dim zodiacal constellation, visible in the northern summer and fall sky. Nevertheless, it is one of the oldest-recognized constellations in the sky, dating back to Sumerian times, thousands of years ago. The star Alpha Capricorni, or α Cap, is a beautiful naked-eye binary.

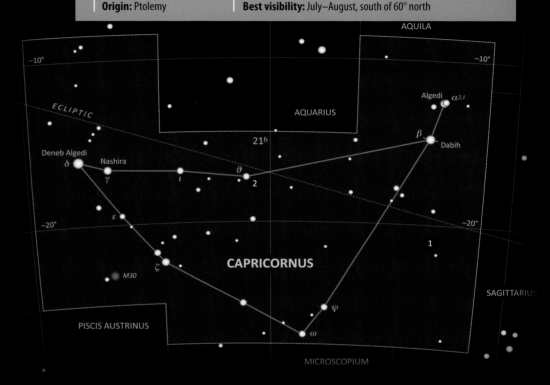

◀ The globular cluster M30 is 28,000 light-years away in the constellation Capricornus.

▼ NASA's Cassini spacecraft captured this image of a giant storm in Saturn's northern hemisphere in 2011. It is similar to the storm observed by Will Hay in 1933, when Saturn was in Capricornus.

▲ The first-ever telescopic sketch of the moon, made by Galileo Galilei on November 30, 1609, when the lunar crescent was in Capricornus.

TIMELINE

1609 ❶ Galileo Galilei aims his newly built telescope in the direction of Capricornus to make the first detailed telescopic sketches of the crescent moon in history.

1764 Charles Messier discovers the globular cluster M30 in the southeastern part of Capricornus.

1933 ❷ British comedian and amateur astronomer Will Hay discovers a large storm in the atmosphere of Saturn, which is located in Capricornus that year.

CARINA

PASSPORT

Latin name: Carina	**Area:** 494.2 square degrees
English name: Keel	**Number of naked-eye stars:** 225
Genitive: Carinae	**Bordering constellations:** Puppis, Pictor, Volans, Chamaeleon, Musca, Centaurus, Vela
Abbreviation: Car	
Origin: de Lacaille	**Best visibility:** January–February, south of 40° north

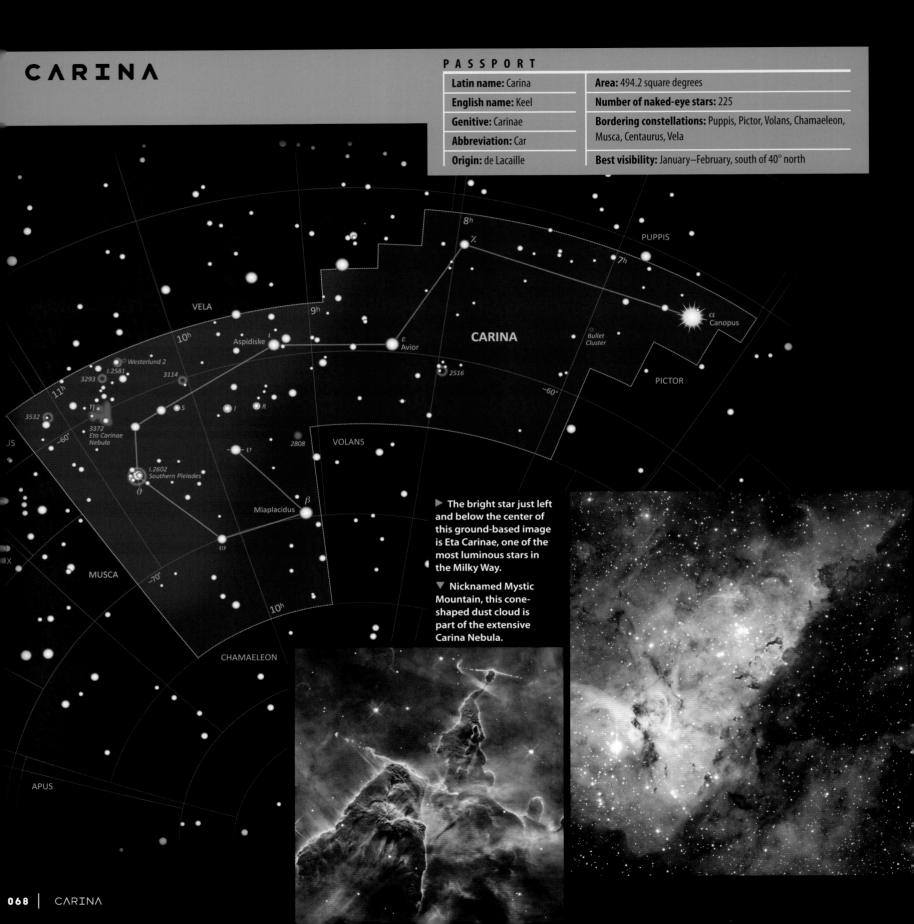

▶ The bright star just left and below the center of this ground-based image is Eta Carinae, one of the most luminous stars in the Milky Way.

▼ Nicknamed Mystic Mountain, this cone-shaped dust cloud is part of the extensive Carina Nebula.

▲ At a distance of 20,000 light-years in Carina is the star-forming region Westerlund 2, imaged by the Hubble Space Telescope on the occasion of its twenty-fifth birthday.

▼ Star cluster NGC 3532 was the target of the first-ever photograph of the Hubble Space Telescope after its launch on April 20, 1990.

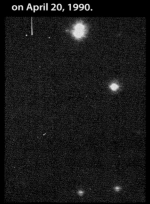

CARINA (THE KEEL) is a huge constellation in the southern sky. Originally, it was part of an even larger constellation, Argo Navis (the Ship *Argo*). Carina is not visible from latitudes north of 40 degrees, but it's a prominent sight for southern-hemisphere observers, even though it lacks a very recognizable shape.

The constellation's brightest star (Alpha Carinae, or α Car) is called Canopus, possibly after the pilot on the Spartan ship during the Trojan War. After Sirius in Canis Major (the Greater Dog), it is the brightest star in the night sky. In fact, Canopus is much more luminous than nearby Sirius, but appears a bit fainter because of its distance of 310 light-years.

Thanks to its large size and proximity to the central plane of the Milky Way, Carina contains numerous star clusters and nebulae, of which the eye-catching Carina Nebula is by far the most famous.

▼ Blue colors reveal the distribution of dark matter in this optical/X-ray composite image of the Bullet Cluster; pink denotes hot gas.

TIMELINE

1751 From South Africa, French astronomer Nicolas Louis de Lacaille is the first to describe the glowing Eta Carinae Nebula (the Carina Nebula, NGC 3372)—a 300-light-year-wide star-forming region some 8,000 light-years away. The giant nebula has been imaged in detail by the Hubble Space Telescope.

1763 De Lacaille creates Carina by dividing Argo Navis up into three smaller constellations. The other two are Puppis (the Stern) and Vela (the Sails).

1843 During a spectacular outburst, the giant star Eta Carinae, or η Car—one of the most luminous stars in the Milky Way galaxy—temporarily becomes as bright as Sirius, the brightest star in the night sky.

1990 On May 20, the open star cluster NGC 3532 in Carina is the first object to be imaged by the newly launched Hubble Space Telescope.

2004 X-ray observations of the Bullet Cluster of galaxies (actually two colliding clusters, also known as 1E 0657-558), at 4 billion light-years in Carina, support the theory that the universe is dominated by mysterious dark matter.

2015 To celebrate its twenty-fifth birthday, the Hubble Space Telescope captures an image of the Westerlund 2 star cluster and its surrounding nebulosity—a large stellar cradle in Carina.

CASSIOPEIA

PASSPORT

Latin name: Cassiopeia		**Area:** 598.4 square degrees	
English name: Cassiopeia		**Number of naked-eye stars:** 157	
Genitive: Cassiopeiae		**Bordering constellations:** Cepheus, Lacerta, Andromeda, Perseus, Camelopardalis	
Abbreviation: Cas			
Origin: Ptolemy		**Best visibility:** September–October, north of 10° south	

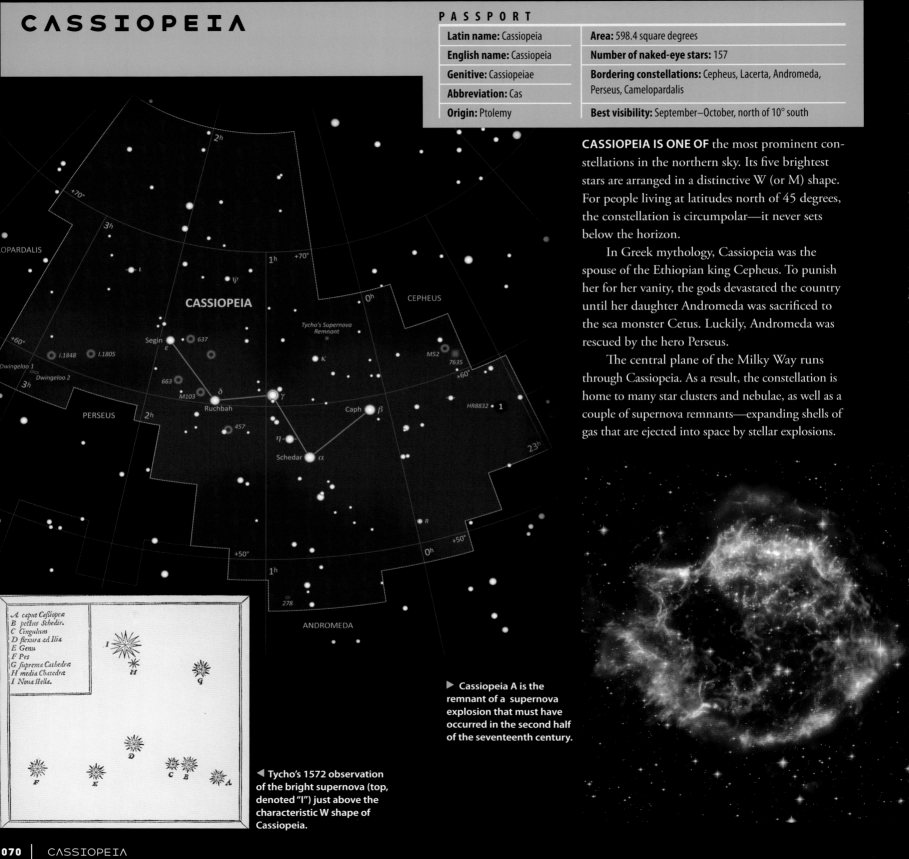

CASSIOPEIA IS ONE OF the most prominent constellations in the northern sky. Its five brightest stars are arranged in a distinctive W (or M) shape. For people living at latitudes north of 45 degrees, the constellation is circumpolar—it never sets below the horizon.

In Greek mythology, Cassiopeia was the spouse of the Ethiopian king Cepheus. To punish her for her vanity, the gods devastated the country until her daughter Andromeda was sacrificed to the sea monster Cetus. Luckily, Andromeda was rescued by the hero Perseus.

The central plane of the Milky Way runs through Cassiopeia. As a result, the constellation is home to many star clusters and nebulae, as well as a couple of supernova remnants—expanding shells of gas that are ejected into space by stellar explosions.

▶ **Cassiopeia A is the remnant of a supernova explosion that must have occurred in the second half of the seventeenth century.**

◀ **Tycho's 1572 observation of the bright supernova (top, denoted "I") just above the characteristic W shape of Cassiopeia.**

TIMELINE

1572   Danish astronomer Tycho Brahe discovers a supernova in Cassiopeia, bright enough to be visible during the day. Tycho's Supernova Remnant is still visible at the same location in the sky.

1680   ❶ On August 16, Royal Astronomer John Flamsteed observes a star that he lists as 3 Cassiopeiae, but that has never been seen since. It might have been the heavily dust-obscured supernova that produced the Cassiopeia A supernova remnant, which has an estimated age of just over three hundred years.

1854   An early French science fiction author, Charlemagne Defontenay, publishes his novel *Star, ou Psi de Cassiopeé*, in which Earth receives a message from tall, immortal aliens living on a planet orbiting the star Psi Cassiopeia, or ψ Cas.

1994   Using their 82-foot radio telescope in Dwingeloo, Dutch radio astronomers discover the galaxies Dwingeloo 1 and Dwingeloo 2, at a distance of some 10 million light-years. They had escaped earlier detection because their light is heavily obscured by dust in our own Milky Way galaxy.

2015   A small, rocky planet is found orbiting the star HR 8832 at 21 light-years' distance in Cassiopeia—the nearest rocky exoplanet to our solar system at that time. Later, four and maybe even six more planets are discovered.

▲ NGC 278 is a galaxy in Cassiopeia at a distance of 38 million light-years, undergoing a burst of star formation.

◀ The obscured spiral galaxy Dwingeloo I just faintly shows up in this photograph made with the Isaac Newton Telescope at La Palma.

▶ The star Kappa Cassiopeia is speeding through the Milky Way at 1,100 kilometers per second, producing a bow shock that shows up in red in this infrared image.

CENTAURUS

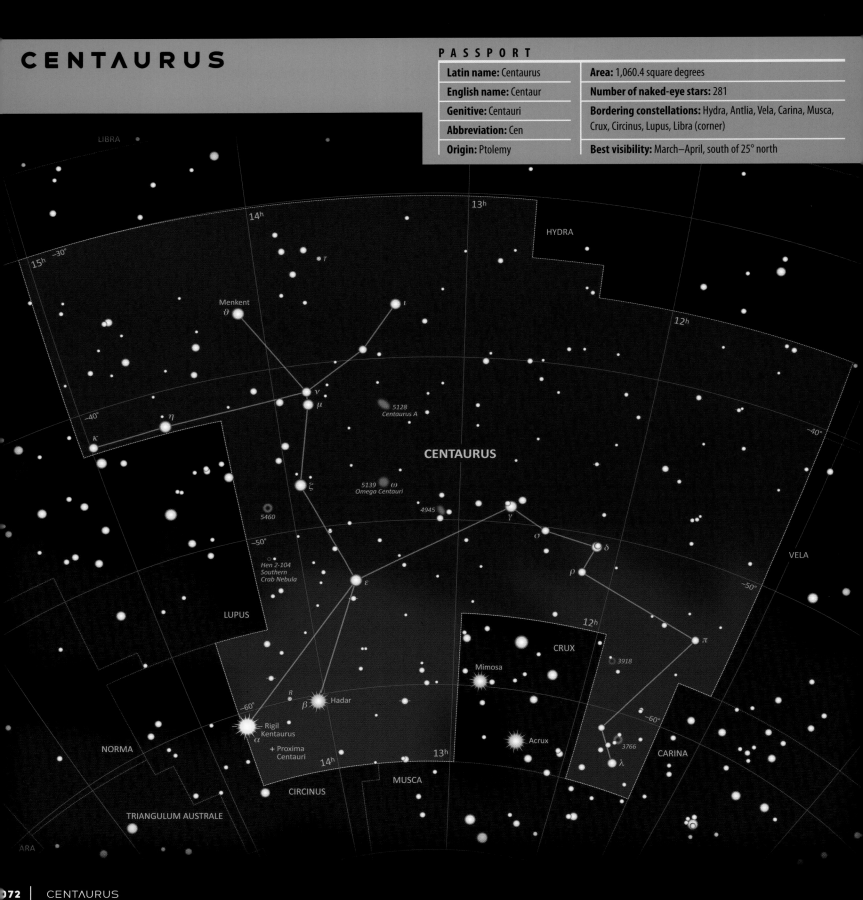

▼ The Centaur appears (mirror-reversed) on a 1602 celestial globe by Willem Blaeu.

CENTAURUS (THE CENTAUR) is a large and prominent constellation in the southern sky. Unfortunately, it can't be observed from North America or Europe. The famous Southern Cross used to be part of Centaurus—sitting between the legs of the mythological creature—but is now regarded as a separate constellation. In Greek mythology, Centaurus represented Chiron, the wise half man half horse. Incidentally, the zodiacal constellation Sagittarius (the Archer) is yet another celestial centaur.

Centaurus is best known for its brightest star, Alpha Centauri or α Cen, also known as Rigil Kentaurus. At a distance of just 4.4 light-years (some 26 trillion miles), the triple star system is the nearest neighbor of our own sun. The faint companion Proxima Centauri is even a bit closer, at 4.25 light-years.

Centaurus also contains two famous deep-sky objects: the bright globular cluster Omega Centauri, or ω Cen—easily visible to the naked eye—and the relatively close radio galaxy Centaurus A (NGC 5128), at some 11 million light-years from Earth.

▼ Hypothetical view from the surface of Proxima b, the nearest planet outside our own solar system, orbiting the red dwarf star Proxima Centauri. The binary star Alpha Centauri can be seen to the upper right of Proxima Centauri.

▲ Proxima Centauri, our nearest stellar neighbor, as seen by the Hubble Space Telescope's Wide Field and Planetary Camera 2.

▶ Nicknamed the Southern Crab because of its remarkable shape, Henize 2-104 is a gaseous nebula in the constellation Centaurus, produced by a binary star in the very center.

▲ NGC 5128 is a huge galaxy, dissected by a turbulent band of dark dust. It harbors a supermassive black hole, and is also a bright source of radio waves, known as Centaurus A.

▼ Omega Centauri is the brightest globular cluster in the night sky. It contains millions of individual stars.

1677 Observing from the island of St. Helena in the South Atlantic Ocean, English astronomer Edmund Halley notices the fuzziness of Omega Centauri, suggesting it is a nebula instead.

1689 French astronomer and Jesuit priest Jean Richaud discovers the binary nature of Alpha Centauri using a telescope in Pondicherry, India. It was one of the very first binary stars to be discovered.

1826 Scottish astronomer James Dunlop is the first to identify Omega Centauri as a globular cluster. It is actually the brightest globular cluster in the night sky. It may be the remaining core of a dwarf galaxy that has been swallowed and largely disrupted by tidal forces from our Milky Way galaxy.

1826 James Dunlop is also the first to discover the galaxy NGC 5128. The galaxy is reported to be just visible to the naked eye under ideal circumstances.

1833 For the first time ever, the distance of another star is measured through the parallax method. However, Thomas Henderson doesn't publish his results on Alpha Centauri until six years later, soon after Friedrich Bessel reports similar observations of the star 61 Cygni.

1847 John Herschel observes the dark dust band that appears to cut the galaxy NGC 5128 in two.

1915 Scottish astronomer Robert Innes discovers Proxima Centauri— a faint red dwarf star that turns out to be the third (and closest) member of the Alpha Centauri system, orbiting the binary star once every 500,000 years or so.

1935 Dutch astronomer Joan Voûte finds that Beta Centauri, or β Cen (also known as Hadar), is also a binary star. In the night sky, Beta Centauri appears almost as bright as Alpha Centauri, but it is much farther away, at some 390 light-years.

1949 The bright radio source Centaurus A is found to be coincident with the galaxy NGC 5128, which probably harbors a supermassive black hole, spewing out energetic jets of particles and radiation into space.

2016 Russian billionaire Yuri Milner and British physicist Stephen Hawking announce the Breakthrough Starshot project, with the ultimate goal of sending swarms of miniature spacecraft to Alpha Centauri. Even though the nanocraft would be accelerated to 20 percent of the speed of light, it would take them more than twenty years to reach their destination.

2016 A more-or-less Earth-like exoplanet is found orbiting Proxima Centauri every 11.2 days at a distance of just 4.6 million miles, in the so-called habitable zone of the faint red dwarf star. Proxima b is the nearest planet outside our own solar system. However, given the star's lethal X-ray outbursts, the prospects for life on the planet are slim.

▲ At the dense core of the globular cluster Omega Centauri, the Hubble Space Telescope reveals tens of thousands of old stars.

▶ Even a small telescope reveals the binary nature of Alpha Centauri. This image was captured by the Hubble Space Telescope.

CEPHEUS

PASSPORT

Latin name: Cepheus	**Area:** 587.8 square degrees
English name: Cepheus	**Number of naked-eye stars:** 152
Genitive: Cephei	**Bordering constellations:** Ursa Minor, Draco, Cygnus, Lacerta, Cassiopeia, Camelopardalis
Abbreviation: Cep	
Origin: Ptolemy	**Best visibility:** July–September, north of the equator

CEPHEUS IS A CONSTELLATION in the northern sky, located in the triangular area between the Pole Star, the star Deneb in the constellation Cygnus (the Swan), and the W-shaped constellation Cassiopeia. It doesn't contain particularly bright stars, but it has a characteristic shape, similar to a house in a children's drawing.

In Greek mythology, Cepheus was the king of Ethiopia, husband of queen Cassiopeia, and father of Andromeda. The band of the Milky Way runs through the southern part of the constellation; as a result, Cepheus contains many nebulae and star clusters.

The most famous object in the constellation, however, is the star Delta Cephei, or δ Cep. It is a variable star and the prototype of Cepheids, which play an important role in establishing the distance scale of the universe.

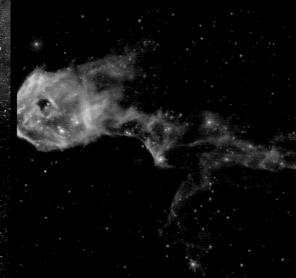

▲ NASA's infrared Spitzer Space Telescope reveals protostars in the Elephant's Trunk Nebula in Cepheus—a so-called globule that appears as a dark cloud on optical photographs.

▲ The bright star at the upper left of the nebula IC 1396 is Herschel's Garnet Star, Mu Cephei. The Elephant's Trunk Nebula is on the upper right.

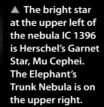

◄ Illustration of a dust-obscured active galactic nucleus like S5 0014+81 in Cepheus, which contains one of the most massive black holes known.

TIMELINE

1783 William Herschel is the first one to note the extreme red color of the star Mu Cephei, or μ Cep. Nicknamed the Garnet Star, it is a distant red supergiant, some 1,500 times the size of our own sun and a few hundred thousand times more luminous.

1784 At the age of twenty, and just two years before his untimely death, deaf and mute English amateur astronomer John Goodricke discovers the variability of the star Delta Cephei, which regularly brightens and dims every 5.37 days.

1912 Henrietta Leavitt shows that variable stars like Delta Cephei—called Cepheids—can be used as distance indicators. A Cepheid's period of variability reveals its true (average) luminosity, and by comparing that value to its observed apparent brightness, its distance can be estimated.

2009 Observations by NASA's Swift spacecraft of the very remote galaxy SS 0014+81 in the northern part of Cepheus indicate that this so-called blazar harbors a central supermassive black hole that weighs in at 40 billion times the mass of the sun—one of the most massive black holes known.

2013 Infrared measurements by the European Herschel Space Observatory reveal very massive protostellar cores in the star-forming region NGC 7538 in Cepheus, more than forty times the mass of the sun.

▶ Star-forming region NGC 7538 in Cepheus, imaged here at infrared wavelengths, contains dozens of very massive protostars.

▶ The plaque in York, England, commemorating John Goodricke's observations of the variability Delta Cephei.

From a window in Treasurers House near this tablet, the young deaf and dumb astronomer
JOHN GOODRICKE
1764 - 1786
who was elected a Fellow of the Royal Society at the age of 21, observed the periodicity of the star ALGOL and discovered the variation of δ CEPHEI and other stars thus laying the foundation of modern measurement of the Universe

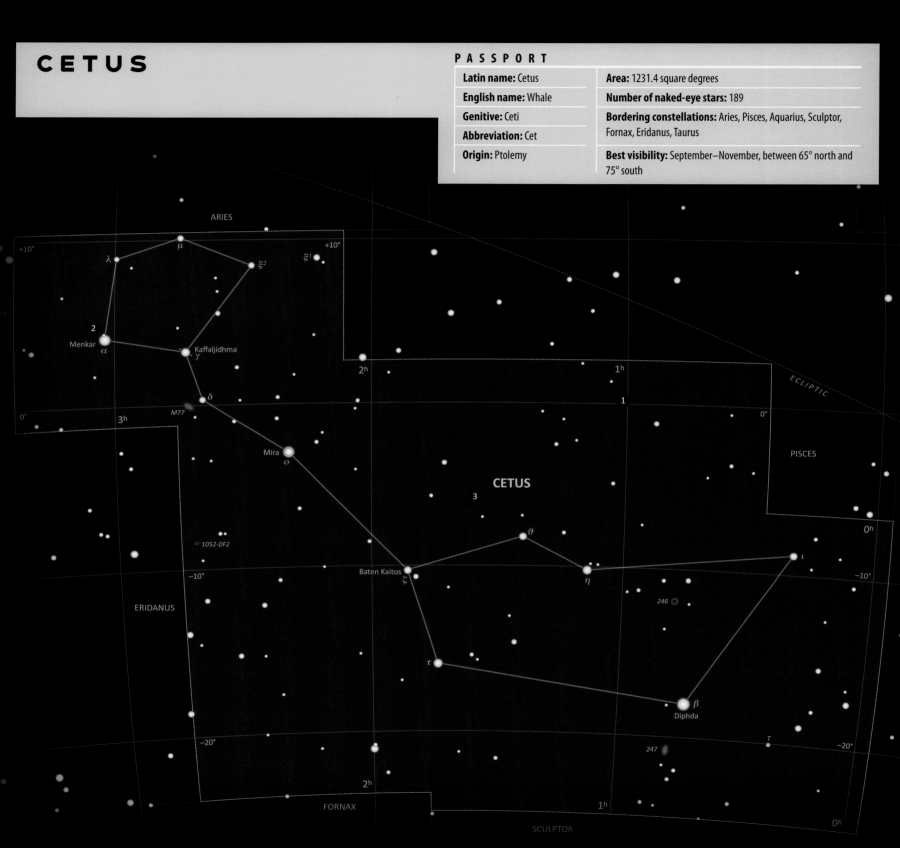

CETUS

PASSPORT

Latin name: Cetus

English name: Whale

Genitive: Ceti

Abbreviation: Cet

Origin: Ptolemy

Area: 1231.4 square degrees

Number of naked-eye stars: 189

Bordering constellations: Aries, Pisces, Aquarius, Sculptor, Fornax, Eridanus, Taurus

Best visibility: September–November, between 65° north and 75° south

ARIES

+10°   +10°

μ

λ

π₂

π₁

2

Menkar
α   Kaffaljidhma
               γ

δ   1

0°   3h   M77   0°

PISCES

Mira
ο

CETUS

3

ϑ

○ 1052-DF2   ι

Baten Kaitos   η   −10°
−10°   ζ

246 ○

ERIDANUS

τ

β
Diphda

−20°   τ   −20°

247

2h

FORNAX   1h   0h

SCULPTOR

CETUS (THE WHALE) is a large constellation straddling the celestial equator. It contains a number of pretty bright stars, but it lacks a very conspicuous shape, making it hard to recognize in the sky. Cetus is also known as the Sea Monster—in Greek mythology, it represented the heinous creature to which Andromeda, daughter of Queen Cassiopeia and King Cepheus of Ethiopia, had to be sacrificed.

Cetus contains the famous star Mira (Omicron Ceti, or o Cet), the very first variable star to be recognized by European astronomers. Another famous object in Cetus is the galaxy M77.

Although it is not part of the zodiac, Cetus is close to the ecliptic—the band on the sky in which the sun, the moon, and the major planets can be found. Many minor planets and Kuiper Belt objects have been discovered in this constellation.

▲ The Seyfert galaxy M77 harbors a supermassive black hole.

▲ The dwarf galaxy NGC 1052-DF2 in Cetus appears to be completely devoid of dark matter.

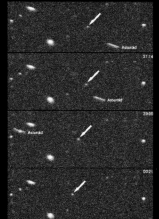

▶ Discovery images of the first Kuiper Belt Object, Albion (1992 QB1, arrowed).

◄ While moving through space, Mira (Omicron Ceti, far right) leaves a comet-like tail of tenuous, hot gas behind, as seen in this ultraviolet image captured by NASA's Galex spacecraft.

TIMELINE

1596 German amateur astronomer David Fabricius is the first to note the variable brightness of the star Omicron Ceti, later to be named Mira ("the miraculous"). At a distance of 420 light-years, Mira slowly varies in size and luminosity every 332 days. Most of the time it is invisible to the naked eye.

1780 French astronomer Pierre Méchain discovers the galaxy M77, close to the star Delta Ceti or δ Cet. Because of its energetic nucleus, it is classified as a so-called Seyfert galaxy.

1960 As part of Project Ozma, U.S. radio astronomer Frank Drake aims a radio telescope at the nearby sun-like star Tau Ceti, or τ Cet, in the hope of intercepting alien messages. In 2012, Tau Ceti is found to be accompanied by a system of at least four planets.

1992 ❶ David Jewitt and Jane Luu find the first Kuiper Belt object other than Pluto in Cetus. Originally known as 1992 QB1, the icy object, orbiting the sun beyond Neptune, is officially named Albion in 2018.

2003 ❷ A team led by Mike Brown discovers the mysterious object Sedna in Cetus. Sedna orbits the sun every 11,500 years or so in an extremely elongated path, which never comes close to Neptune's orbit.

2005 ❸ Cetus is also the constellation in which Brown and his colleagues discover the dwarf planet Eris. Although Eris is slightly smaller than Pluto, it is 25 percent more massive.

2017 By studying the spatial motions of planetary nebulae in the dwarf galaxy NGC 1052-DF2 in Cetus, American astronomers discover that the galaxy appears to be completely devoid of dark matter—an observation that is hard to explain by alternative theories of gravity.

◄ Artist's impression of Sedna, with the sun in the distance.

▶ Artist's impression of the bright and massive dwarf planet Eris with its large moon, Dysnomia.

CHAMΛELEON

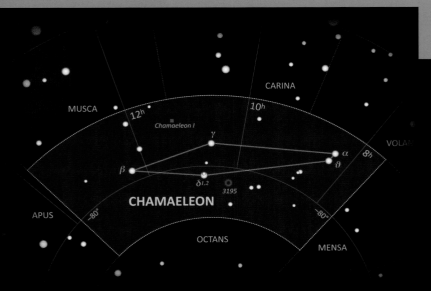

MUSCA · 12ʰ · Chamaeleon I · CARINA · 10ʰ · γ · VOLAN · α · ϑ · 8ʰ · β · δ1.2 · 3195 · -80° · -80° · **CHAMAELEON** · APUS · OCTANS · MENSA

PASSPORT

Latin name: Chamaeleon	**Area:** 131.6 square degrees
English name: Chameleon	**Number of naked-eye stars:** 31
Genitive: Chamaeleontis	**Bordering constellations:** Musca, Carina, Volans, Mensa, Octans, Apus
Abbreviation: Cha	
Origin: Keyser and de Houtman	**Best visibility:** February–March, south of 5° north

▶ **A very young protostar in the Chamaeleon I star-forming region ejects material into space. The star itself is hidden by dust.**

◀ **Cold dust particles reflect the bluish light from the young star HD 97300, embedded in the Chamaeleon star-forming region.**

CHAMAELEON (THE CHAMELEON) is a hard-to-recognize constellation close to the south celestial pole. It contains a huge star-forming region, at a distance of just 500 light-years, consisting of three distinct molecular clouds, Chamaeleon I, II, and III (or Cha I, Cha II, and Cha III).

TIMELINE

1598 — Based on observations by Dutch seafarers Pieter Dirkszoon Keyser and Frederick de Houtman, Flemish cartographer Petrus Plancius introduces Chamaeleon as a new constellation.

1835 — John Herschel discovers planetary nebula NGC 3195 in Chamaeleon.

1991 — The German X-ray satellite ROSAT discovers dozens of X-ray sources in the Chamaeleon I star-forming region, most of them young stellar objects.

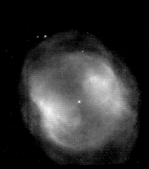

◀ **NGC 3195 is a faint planetary nebula in Chamaeleon, discovered by John Herschel.**

CIRCINUS

PASSPORT

Latin name: Circinus	**Area:** 93.4 square degrees
English name: Draftman's Compasses	**Number of naked-eye stars:** 39
Genitive: Circini	**Bordering constellations:** Lupus, Centaurus, Musca, Apus, Triangulum Australe, Norma
Abbreviation: Cir	**Best visibility:** April–May, south of 15° north
Origin: de Lacaille	

◄ The active Circinus galaxy (ESO 97-G13), as imaged by the Hubble Space Telescope.

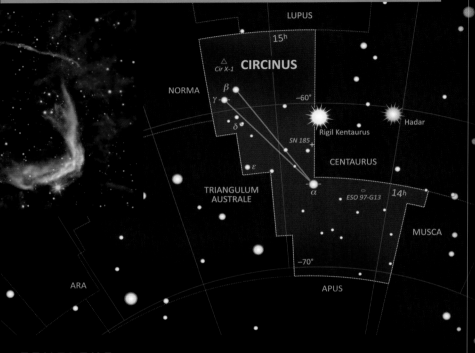

CIRCINUS (THE DRAFTMAN'S COMPASSES) is a small constellation in the southern sky. It is easy to find, though, since its three brightest stars are close to the bright star Alpha Centauri. In fact, those stars used to be part of the constellation Centaurus (the Centaur).

▲ The glowing nebula RCW 86 in Circinus is most likely the expanding remnant of the supernova of 185 AD.

▼ NASA's Chandra X-ray Observatory has imaged "light echoes" around Circinus X-1: X-rays that bounce off gas clouds between the X-ray source and the Earth.

TIMELINE

185 — Chinese astronomers note the appearance of a bright "guest star" (a supernova) in the part of the sky that is now occupied by the constellation Circinus, very close to the star Alpha Centauri or α Cen. SN 185 is the oldest supernova explosion for which records exist.

1763 — Circinus is introduced as a new constellation by French astronomer Nicolas Louis de Lacaille in his southern star catalog *Coelum Australe Stelliferum*.

1969 — A rocket experiment detects X-rays from a neutron star in a binary system in Circinus. The source, Circinus X-1, is more than 30,000 light-years away.

1977 — Astronomers discover the relatively close but heavily dust-obscured Circinus galaxy (ESO 97-G13), at a distance of just 13 million light-years.

COLUMBA

PASSPORT

Latin name: Columba	**Area:** 270.2 square degrees
English name: Dove	**Number of naked-eye stars:** 68
Genitive: Columbae	**Bordering constellations:** Lepus, Caelum, Pictor, Puppis, Canis Major
Abbreviation: Col	
Origin: Plancius	**Best visibility:** November–January, south of 45° north

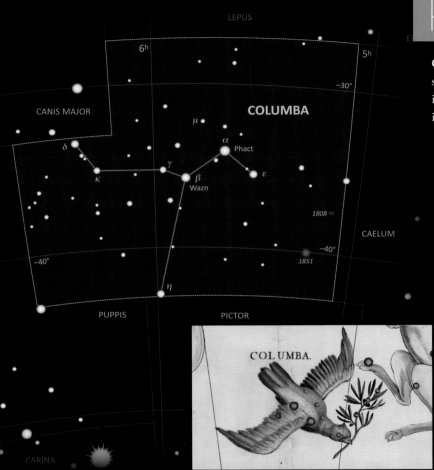

COLUMBA (THE DOVE) is a small but rather conspicuous constellation south of Lepus (the Hare), invented by Dutch sailors as Noah's dove, carrying an olive branch.

▶ At over 100 kilometers per second, Mu Columbae races away from a star-forming region in Orion.

◀ Columba (the Dove) carrying an olive branch. This map is from the 1687 star atlas of Polish astronomer Johannes Hevelius.

▼ NGC 1808, an active galaxy in the westernmost part of the constellation Columba, was home to supernova 1993af.

TIMELINE

1612   Columba appears for the first time as a separate constellation on a celestial globe made by Petrus Plancius.

1783   William Herschel discovers that our solar system, in its orbit around the center of the Milky Way, is moving toward the constellation Hercules and away from the constellation Columba.

1954   Dutch astronomer Adriaan Blaauw identifies the star Mu Columbae, or μ Col, as a runaway star, ejected from a star-forming region in the constellation Orion. The star AE Aurigae is moving away from the same region in the opposite direction.

COMA BERENICES

PASSPORT

Latin name: Coma Berenices		**Area:** 386.5 square degrees	
English name: Berenice's Hair		**Number of naked-eye stars:** 66	
Genitive: Comae Berenices		**Bordering constellations:** Canes Venatici, Ursa Major, Leo, Virgo, Boötes	
Abbreviation: Com			
Origin: Vopel		**Best visibility:** March–April, north of 55° south	

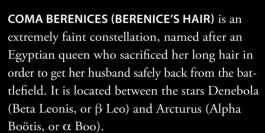

◄ Known as "the Mice" because of their long tidal tails, these two galaxies in Coma Berenices are gravitationally interacting and due to merge in the distant future.

COMA BERENICES (BERENICE'S HAIR) is an extremely faint constellation, named after an Egyptian queen who sacrificed her long hair in order to get her husband safely back from the battlefield. It is located between the stars Denebola (Beta Leonis, or β Leo) and Arcturus (Alpha Boötis, or α Boo).

The constellation is mainly known for the sprawling naked-eye star cluster Melotte 111 and for the large number of remote galaxies it contains. Looking in the direction of Coma Berenices, one looks straight up from the central plane of the Milky Way galaxy, way out into deep space.

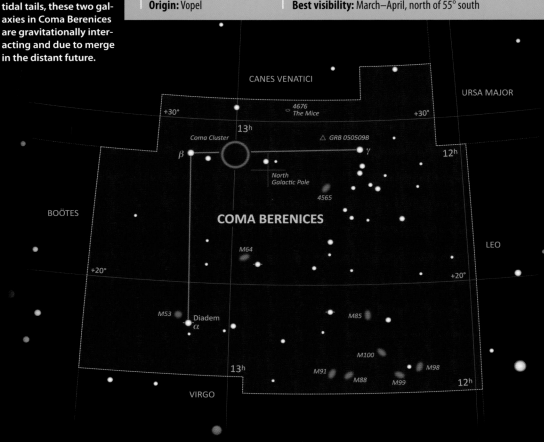

► The V-shaped star cluster Melotte 111 was originally seen as the plume on the tail of Leo (the Lion).

◄ M53 is the brightest globular cluster in Coma Berenices.

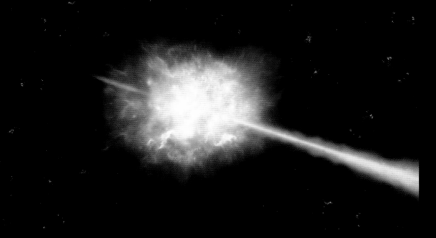

▼ Gamma-ray burst GRB050509B was the energetic explosion caused by the collision and merger of two ultra-compact neutron stars.

TIMELINE

1536 German cartographer Caspar Vopel is the first one to introduce Coma Berenices as a separate constellation.

1775 Johann Bode discovers the globular cluster M53 in Coma Berenices. It contains extremely old stars and is located at almost 60,000 light-years from Earth, in the outskirts of our Milky Way galaxy.

1915 British astronomer Philibert Melotte lists the Coma star cluster—visible to the naked eye as a hazy patch of faint stars—as entry 111 in his new catalogue of star clusters. Melotte 111 contains some fifty stars and lies at a distance of just 280 light-years.

1933 Studying galaxy motions in the Coma cluster of galaxies, Swiss-American astronomer Fritz Zwicky concludes that the cluster must contain large amounts of invisible dark matter. The true nature of dark matter is still a mystery.

2005 The first short gamma-ray burst with a detectable afterglow (GRB 050509B) is observed in Coma Berenices. The discovery supports the theory that short gamma-ray bursts are due to neutron star collisions.

◄ At a distance of 280 million light-years, the Coma cluster contains many thousands of individual galaxies.

CORONA AUSTRALIS

Latin name: Corona Australis	**Area:** 127.7 square degrees
English name: Southern Crown	**Number of naked-eye stars:** 46
Genitive: Coronae Australis	**Bordering constellations:** Sagittarius, Scorpius, Ara, Telescopium
Abbreviation: CrA	
Origin: Ptolemy	**Best visibility:** June–July, south of 45° north

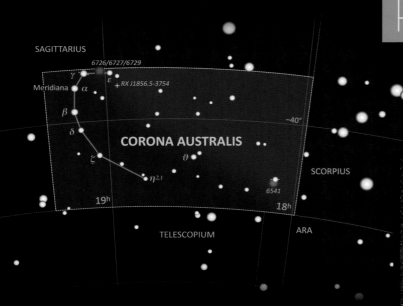

CORONA AUSTRALIS (THE SOUTHERN CROWN) is a faint but nevertheless quite conspicuous arc-shaped constellation south of Sagittarius (the Archer). It contains a large, relatively close star-forming region.

▲ **Dark clouds of dust and bluish reflection nebulae mark one of the star-forming regions in Corona Australis.**

▼ **The Hubble Space Telescope took this image of NGC 6541, the brightest globular cluster in Corona Australis.**

TIMELINE

1826   Italian astronomer Niccolò Cacciatore discovers the globular cluster NGC 6541, which is some 23,000 light-years away, in the southwest corner of Corona Australis.

1861   Julius Schmidt is the first to note three nebulae in Corona Australis (NGC 6726, 6727, and 6729)—the glowing parts of a larger cloud of cold molecular gas from which new stars are born.

1992   Astronomers discover the closest neutron star known, RX J1856.5-3754, at a distance of just 400 light-years from Earth, in the north-eastern part of the constellation.

▶ **Corona Australis appears as a crown in the 1687 star atlas of Johannes Hevelius. In earlier times, it was often depicted as a wreath.**

CORONA BOREALIS

PASSPORT

Latin name: Corona Borealis	**Area:** 178.7 square degrees
English name: Northern Crown	**Number of naked-eye stars:** 37
Genitive: Coronae Borealis	**Bordering constellations:** Hercules, Boötes, Serpens Caput
Abbreviation: CrB	**Best visibility:** April–June, north of 50° south
Origin: Ptolemy	

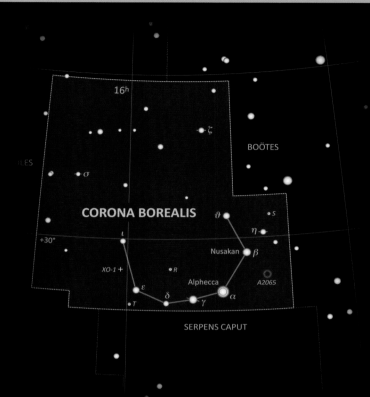

CORONA BOREALIS (THE NORTHERN CROWN) is a small but very conspicuous arc of stars between the constellations Boötes (the Herdsman) and Hercules. It represents the crown of Ariadne, the mythological daughter of King Minos of Crete.

TIMELINE

1795 Edward Pigott discovers the variability of the star R Coronae Borealis, or R CrB—the prototype of a class of variable stars that are surrounded by dusty clouds of carbon.

1866 Another variable star, T Coronae Borealis, or T CrB, normally invisible to the naked eye, temporarily becomes as bright as Alphecca, the brightest star in the constellation.

2006 A team of professional and amateur astronomers discover a Jupiter-sized planet (XO-1b) orbiting a faint star in Corona Borealis.

◀ **R Coronae Borealis** is a yellow supergiant star that fades irregularly when shrouded by ejected clouds of gas and dust.

▶ **Abell 2065** is a cluster of some 400 galaxies at more than a billion light-years from Earth, in the constellation Corona Borealis.

◀ **Exoplanet XO-1b** (right), compared to the planet Jupiter in our own solar system.

CORVUS

PASSPORT

Latin name: Corvus	**Area:** 183.8 square degrees
English name: Raven	**Number of naked-eye stars:** 29
Genitive: Corvi	**Bordering constellations:** Virgo, Crater, Hydra
Abbreviation: Crv	**Best visibility:** March–April, south of 60° north
Origin: Ptolemy	

CORVUS (THE RAVEN) is a small but eye-catching constellation, southwest of the bright star Spica in Virgo (the Virgin). It is home to the most famous pair of interacting galaxies, the Antennae.

◄ A long-exposure photograph reveals the curved tidal tails of the interacting Antennae galaxies.

▼ Bright star clusters light up in the aftermath of the collision of galaxies NGC 4038 and NGC 4039, also known as the Antennae.

TIMELINE

1785 English astronomer William Herschel discovers the interacting galaxies NGC 4038 and NGC 4039. Because of their long tidal tails, the pair has become known as the Antennae. A few billion years from now, our own Milky Way galaxy and the nearby Andromeda galaxy will similarly collide and merge.

1974 Ultraviolet radiation from the star 31 Crateris or 31 Crt, which is actually in the constellation Corvus, is briefly interpreted by NASA's planetary explorer Mariner 10 as coming from a moon of the planet Mercury. In fact, Mercury has no moons.

▼ The constellations Corvus (the Raven, right) and Crater (the Cup, left) appear in mirror-reversed orientation on the 1551 celestial globe of Flemish cartographer Gerardus Mercator.

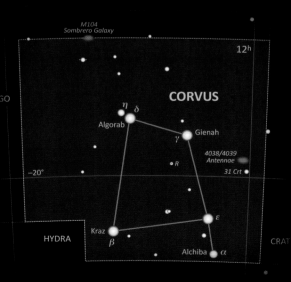

M104 Sombrero Galaxy

VIRGO

12ʰ

CORVUS

η δ

Algorab

Gienah

γ

4038/4039 Antennae

R

−20°

31 Crt

ε

HYDRA

Kraz

β

CRATE

Alchiba α

CRATER

PASSPORT

Latin name: Crater	**Area:** 282.4 square degrees
English name: Cup	**Number of naked-eye stars:** 33
Genitive: Crateris	**Bordering constellations:** Leo, Sextans, Hydra, Corvus, Virgo
Abbreviation: Crt	**Best visibility:** February–April, south of 60° north
Origin: Ptolemy	

CRATER (THE CUP) is larger but less conspicuous than neighboring Corvus (the Raven). In Greek mythology, both constellations are associated with Apollo, the sun god. They sit on the back of Hydra (the Sea Serpent).

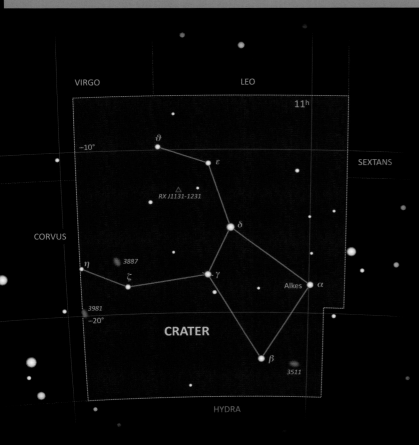

▶ **The gravity of a foreground galaxy produces multiple images of the remote quasar RX J1131-1231, which harbors a rapidly rotating supermassive black hole.**

◀ **Galaxy NGC 3511 in Crater, as imaged by the Carnegie-Irvine Galaxy Survey.**

TIMELINE

1785   Using his home-built telescopes, English astronomer William Herschel discovers a number of galaxies in Crater, including NGC 3887 and NGC 3981. A year later, he also finds NGC 3511.

2014   Observations with NASA's Chandra X-ray Observatory reveal that the supermassive black hole RX J1131-1231 in Crater, located in the core of a quasar at a distance of 6 billion light-years from Earth, rotates at about half the speed of light. It's the first time that astronomers are able to measure the spin rate of a black hole.

CRUX

PASSPORT

Latin name: Crux	**Area:** 68.4 square degrees
English name: Southern Cross	**Number of naked-eye stars:** 49
Genitive: Crucis	**Bordering constellations:** Centaurus, Musca
Abbreviation: Cru	**Best visibility:** March–April, south of 25° north
Origin: Plancius	

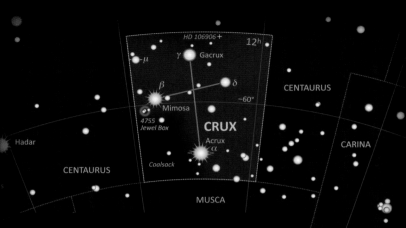

CRUX (THE SOUTHERN CROSS) is one of the most striking constellations in the southern sky, despite being also the smallest. Its compact shape of four bright stars features on the flags of five nations, including Australia, New Zealand, Brazil, Samoa, and Papua New Guinea. Moreover, the constellation serves as a navigational aid: The long axis of the cross points in the direction of the south celestial pole, just like the two "pointers" of the Big Dipper (Ursa Major) help to locate the Pole Star.

Lying smack in the middle of the Milky Way band, Crux contains beautiful star clusters and luminous nebulae, as well as the famous Coalsack Nebula—a dense cloud of dark dust.

▲ The Southern Cross (right) is close to the bright stars Alpha and Beta Centauri (lower left). The Coalsack Nebula is the dark patch adjacent to the kite-shaped constellation.

TIMELINE

1499
Spanish navigator Vicente Yáñez Pinzón is the first to describe the Coalsack Nebula, a dark cloud at a distance of some 600 light-years, measuring 35 light-years across. To Australian Aboriginals, the Coalsack Nebula represents the head of the Giant Emu in the Sky— one of their "dark constellations."

1612
Flemish cartographer Petrus Plancius introduces Crux as a separate constellation. To the Greeks, it was just part of the constellation Centaurus (the Centaur).

1751
During his stay in South Africa, French astronomer Nicolas Louis de Lacaille discovers the Kappa Crucis, or κ Cru, star cluster, a very compact group of young stars some 6,440 light-years from Earth. Later, John Herschel gives the cluster its common nickname, the Jewel Box.

2014
Astronomers discover a planet around the star HD 106906 in Crux, at some 300 light-years' distance from Earth. HD 106906b, as it is called, is eleven times more massive than Jupiter and has the widest orbit of any exoplanet, at almost 62.1 billion miles from its parent star— more than twenty times the distance from the sun to Neptune.

▲ The Southern Cross (right) and Alpha Centauri (left)

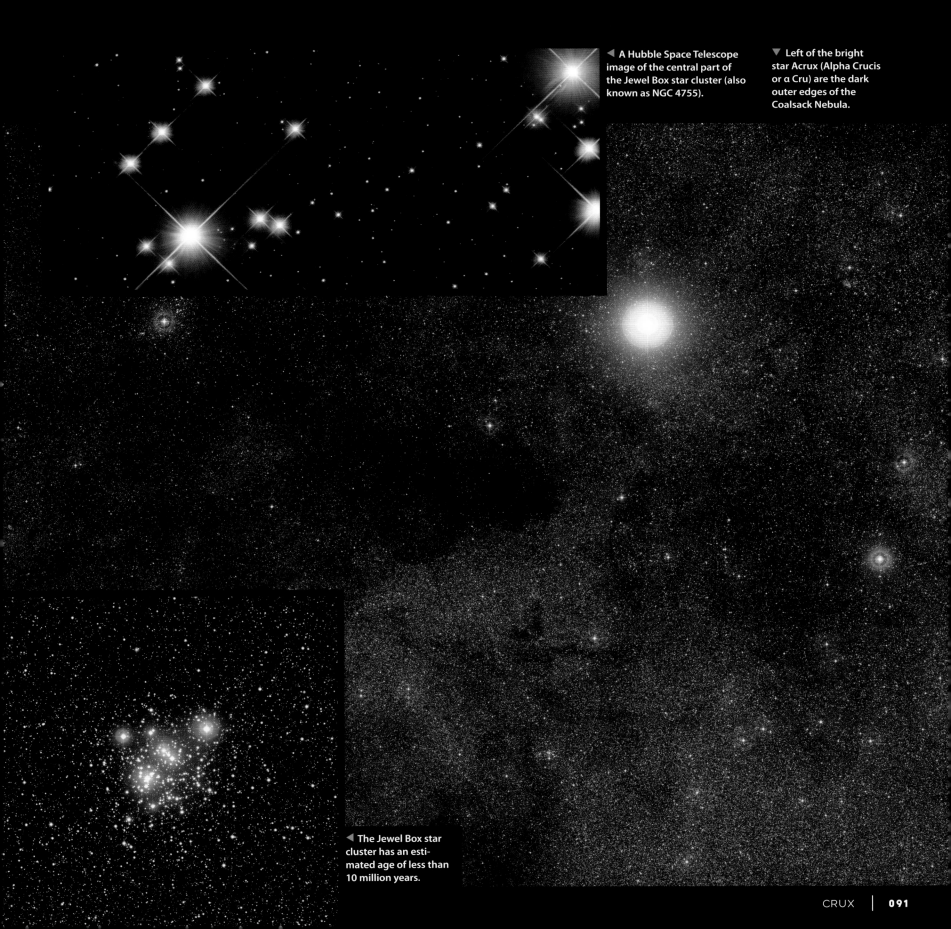

◀ A Hubble Space Telescope image of the central part of the Jewel Box star cluster (also known as NGC 4755).

▼ Left of the bright star Acrux (Alpha Crucis or α Cru) are the dark outer edges of the Coalsack Nebula.

◀ The Jewel Box star cluster has an estimated age of less than 10 million years.

CYGNUS

PASSPORT

Latin name: Cygnus	**Area:** 804.0 square degrees
English name: Swan	**Number of naked-eye stars:** 262
Genitive: Cygni	**Bordering constellations:** Cepheus, Draco, Lyra, Vulpecula, Pegasus, Lacerta
Abbreviation: Cyg	
Origin: Ptolemy	**Best visibility:** June–August, north of 25° south

CYGNUS (THE SWAN) is one of the most striking constellations in the northern sky. Because of its shape, it's sometimes called the Northern Cross, and it really resembles a swan flying overhead. The brightest star in the constellation, Deneb (Alpha Cygni, or α Cyg), marks one of the vertices of the giant Summer Triangle. In Greek mythology, Zeus disguised himself as a swan to seduce Leda, the queen of Sparta. However, the constellation has also been associated with other legends.

The band of the Milky Way runs through Cygnus. As a result, the constellation contains a large number of star clusters and glowing nebulae. Moreover, lots of exoplanets—planets orbiting stars other than our own sun—have been discovered in Cygnus, since NASA's Kepler Space Telescope was aimed at this part of the sky for three years on end.

▲ **The Veil Nebula in Cygnus is part of an old supernova remnant. The bright star in the photo (52 Cyg) is not associated with the nebula.**

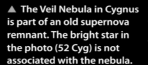

◄ **Amateur shot of the North America Nebula. The bright star at the left is Xi Cygni or ξ Cyg.**

▼ **The erratic brightness variations of Tabby's Star may be caused by an uneven ring of dust and comets surrounding the star.**

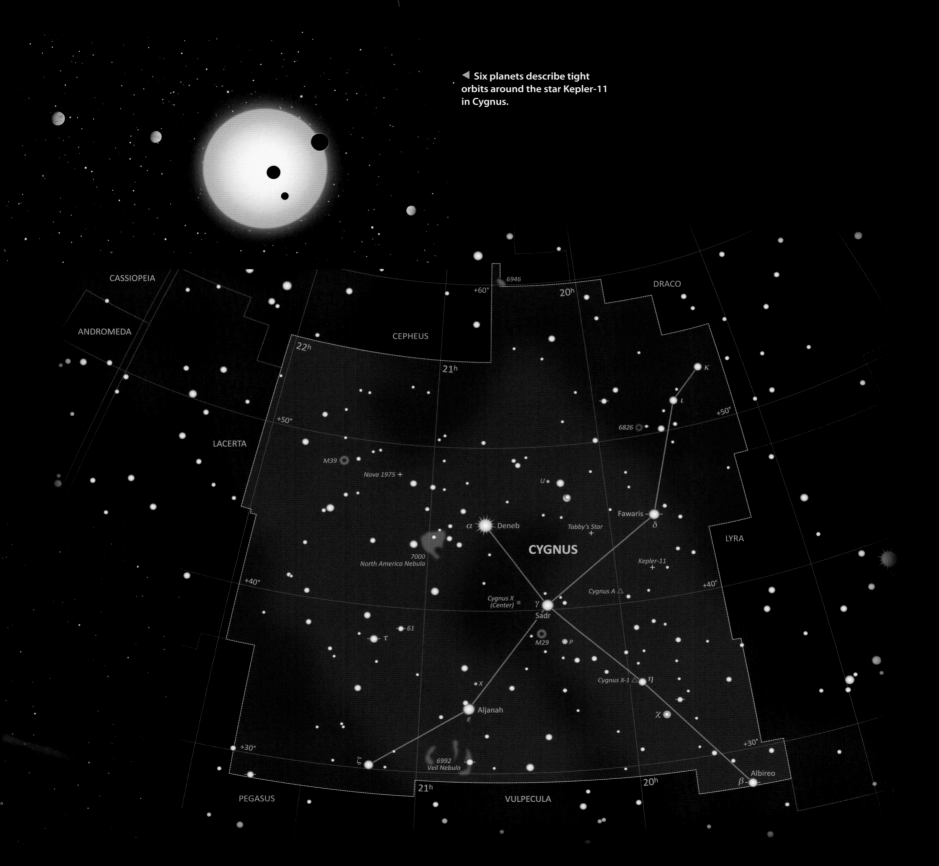

◀ **Six planets describe tight orbits around the star Kepler-11 in Cygnus.**

CASSIOPEIA

ANDROMEDA

LACERTA

CEPHEUS

DRACO

6946

+60°

20ʰ

22ʰ

21ʰ

+50°

+50°

κ

ι

6826

M39

Nova 1975 +

U

Fawaris

δ

LYRA

α Deneb

Tabby's Star
+

CYGNUS

Kepler-11
+

7000
North America Nebula

Cygnus A △

+40°

+40°

Cygnus X
(Center)

γ
Sadr

61

τ

M29

P

χ

Cygnus X-1 △ η

x

Aljanah

ε

χ

+30°

+30°

ξ

6992
Veil Nebula

Albireo

β

21ʰ

20ʰ

PEGASUS

VULPECULA

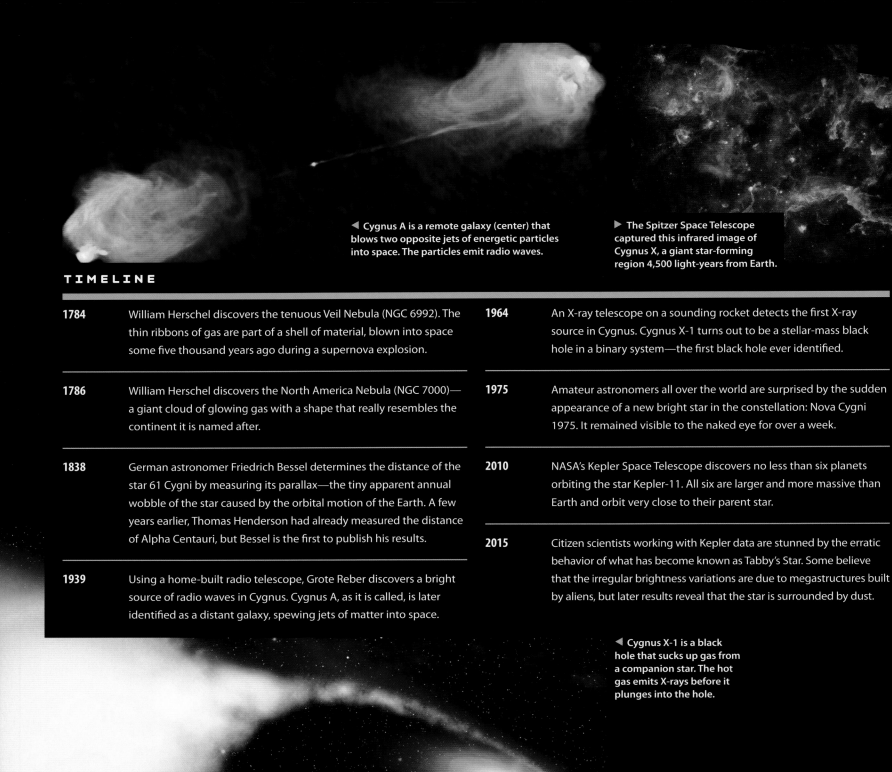

◀ Cygnus A is a remote galaxy (center) that blows two opposite jets of energetic particles into space. The particles emit radio waves.

▶ The Spitzer Space Telescope captured this infrared image of Cygnus X, a giant star-forming region 4,500 light-years from Earth.

TIMELINE

1784 William Herschel discovers the tenuous Veil Nebula (NGC 6992). The thin ribbons of gas are part of a shell of material, blown into space some five thousand years ago during a supernova explosion.

1786 William Herschel discovers the North America Nebula (NGC 7000)— a giant cloud of glowing gas with a shape that really resembles the continent it is named after.

1838 German astronomer Friedrich Bessel determines the distance of the star 61 Cygni by measuring its parallax—the tiny apparent annual wobble of the star caused by the orbital motion of the Earth. A few years earlier, Thomas Henderson had already measured the distance of Alpha Centauri, but Bessel is the first to publish his results.

1939 Using a home-built radio telescope, Grote Reber discovers a bright source of radio waves in Cygnus. Cygnus A, as it is called, is later identified as a distant galaxy, spewing jets of matter into space.

1964 An X-ray telescope on a sounding rocket detects the first X-ray source in Cygnus. Cygnus X-1 turns out to be a stellar-mass black hole in a binary system—the first black hole ever identified.

1975 Amateur astronomers all over the world are surprised by the sudden appearance of a new bright star in the constellation: Nova Cygni 1975. It remained visible to the naked eye for over a week.

2010 NASA's Kepler Space Telescope discovers no less than six planets orbiting the star Kepler-11. All six are larger and more massive than Earth and orbit very close to their parent star.

2015 Citizen scientists working with Kepler data are stunned by the erratic behavior of what has become known as Tabby's Star. Some believe that the irregular brightness variations are due to megastructures built by aliens, but later results reveal that the star is surrounded by dust.

◀ Cygnus X-1 is a black hole that sucks up gas from a companion star. The hot gas emits X-rays before it plunges into the hole.

DELPHINUS

PASSPORT

Latin name: Delphinus	**Area:** 188.5 square degrees
English name: Dolphin	**Number of naked-eye stars:** 44
Genitive: Delphini	**Bordering constellations:** Vulpecula, Sagitta, Aquila, Aquarius, Equuleus, Pegasus
Abbreviation: Del	
Origin: Ptolemy	**Best visibility:** July–August, north of 65° south

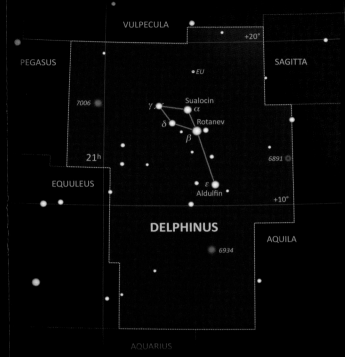

◀ Delphinus appears as a rather monstrous creature on this 1825 star map from *Urania's Mirror*.

▶ NGC 6891 is a small planetary nebula in Delphinus, some 7,000 light-years away.

DELPHINUS (THE DOLPHIN) may be one of the smaller constellations in the sky, but it is easy to recognize: It contains a diamond-shaped group of four stars and is northeast of the bright star Altair in Aquila (the Eagle).

TIMELINE

1784   William Herschel discovers the globular cluster NGC 7006 in Delphinus—a spherical collection of stars approximately 140,000 light-years away.

1814   Niccolò Cacciatore, at the time an assistant at the Observatory of Palermo, Sicily, reverses the Latinized version of his name (Nicolaus Venator) to name the stars Alpha and Beta Delphini: α Del is now known as Sualocin, β Del is called Rotanev.

1884   Scottish astronomer Ralph Copeland discovers the planetary nebula NGC 6891, at the very western edge of the constellation.

▶ At a distance of 140,000 light-years, NGC 7006 in Delphinus is one of the most remote globluar clusters in

DORADO

PASSPORT

Latin name: Dorado	**Area:** 179.2 square degrees
English name: Goldfish	**Number of naked-eye stars:** 29
Genitive: Doradus	**Bordering constellations:** Caelum, Horologium, Reticulum, Hydrus, Mensa, Volans, Pictor
Abbreviation: Dor	
Origin: Keyser and de Houtman	**Best visibility:** November–December, south of 25° north

▲ **NGC 1850, discovered by James Dunlop in 1826, looks like a globular cluster but is much younger.**

▼ **The Tarantula Nebula (30 Doradus) is at the top of this image, captured by the European VST Survey Telescope in Chile.**

DORADO (THE GOLDFISH) is a rather non-descript constellation in the southern sky that wouldn't get much attention if it weren't for the Large Magellanic Cloud (LMC)—a companion galaxy of our Milky Way that lies across the border of Dorado and the neighboring constellation Mensa (the Table Mountain). The Large Magellanic Cloud, named after the Portuguese explorer Fernão de Magalhães (Ferdinand Magellan), contains many dozens of bright star-forming regions, including the giant Tarantula Nebula, and many young star clusters. It was also home to supernova SN 1987A, which was clearly visible to the naked eye, despite its distance from Earth of about 167,000 light-years.

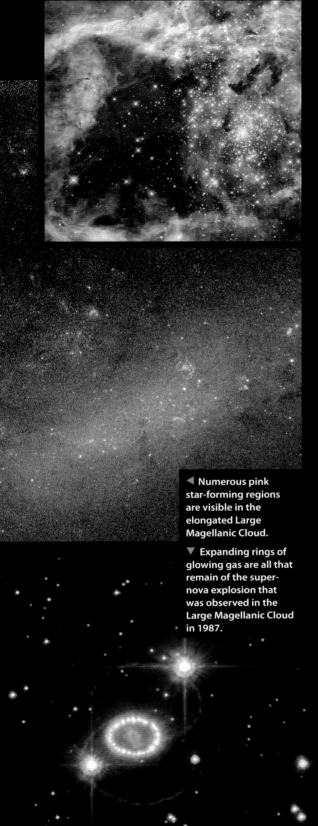

◀ R136 was once thought to be just one star, but is now known to contain many extremely young and massive stellar giants.

TIMELINE

964   Persian astronomer Abd al-Rahman al-Sufi mentions the Large Magellanic Cloud in his *Book of Fixed Stars*. It was barely visible above the horizon from southern Arabia.

1598   Dorado is introduced as a separate constellation by Petrus Plancius, based on descriptions by Dutch seafarers Pieter Dirkszoon Keyser and Frederick de Houtman.

1751   Nicolas Louis de Lacaille is the first to realize that the star 30 Doradus is actually a nebula in the LMC. It is now known as the Tarantula Nebula (NGC 2070) because of its spidery shape.

1826   James Dunlop discovers the open star cluster NGC 1850 in the LMC, which more or less looks like a globular cluster, although it is extremely young, astronomically speaking.

1987   On February 23, light from a supernova explosion in the LMC arrives on Earth, together with a cosmic tsunami of neutrinos—the first time that neutrinos from a supernova explosion are observed. Supernova SN 1987A remains visible with the naked eye for many months.

1992   Observations with the Hubble Space Telescope reveal that the luminous star R136 in the LMC is actually a very tight cluster of extremely massive, luminous stars. The brightest member of the cluster, R136a1, is more than three hundred times as massive and almost 9 million times as luminous as the sun.

2005   Astronomers discover a high-velocity star in Doradus, HE 0437-5439, with a recession velocity of 451 miles per second—high enough to escape the gravity of our Milky Way galaxy.

◀ Numerous pink star-forming regions are visible in the elongated Large Magellanic Cloud.

▼ Expanding rings of glowing gas are all that remain of the supernova explosion that was observed in the Large Magellanic Cloud in 1987.

DRACO

PASSPORT

Latin name: Draco	**Area:** 1,083.0 square degrees
English name: Dragon	**Number of naked-eye stars:** 211
Genitive: Draconis	**Bordering constellations:** Ursa Minor, Camelopardalis, Ursa Major, Boötes, Hercules, Lyra, Cygnus, Cepheus
Abbreviation: Dra	
Origin: Ptolemy	**Best visibility:** April–June, north of the equator

DRACO (THE DRAGON) is a huge constellation in the northern sky. It represents the dragon that was slain by Hercules in one of his twelve labors. The constellation is quite easy to find, thanks to its sinuous shape. The dragon's tail is located between the Big Dipper and the Little Dipper; its compact head lies close to the bright star Vega in Lyra (the Lyre). A few thousand years ago, the star Thuban (Alpha Draconis, or α Dra) was our Pole Star; some twenty-one thousand years from now, Earth's axis will again point in the direction of this star.

Draco contains many binary stars and remote galaxies. It is also home to one of the best-known planetary nebulae, the Cat's Eye Nebula.

▶ **The Draco Dwarf galaxy can be recognized as a slight overdensity of faint stars.**

▲ **The Tadpole galaxy (UGC 10214) in Draco sports an elongated tidal tail, produced by the gravity of a small companion galaxy.**

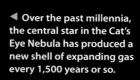

◀ **Over the past millennia, the central star in the Cat's Eye Nebula has produced a new shell of expanding gas every 1,500 years or so.**

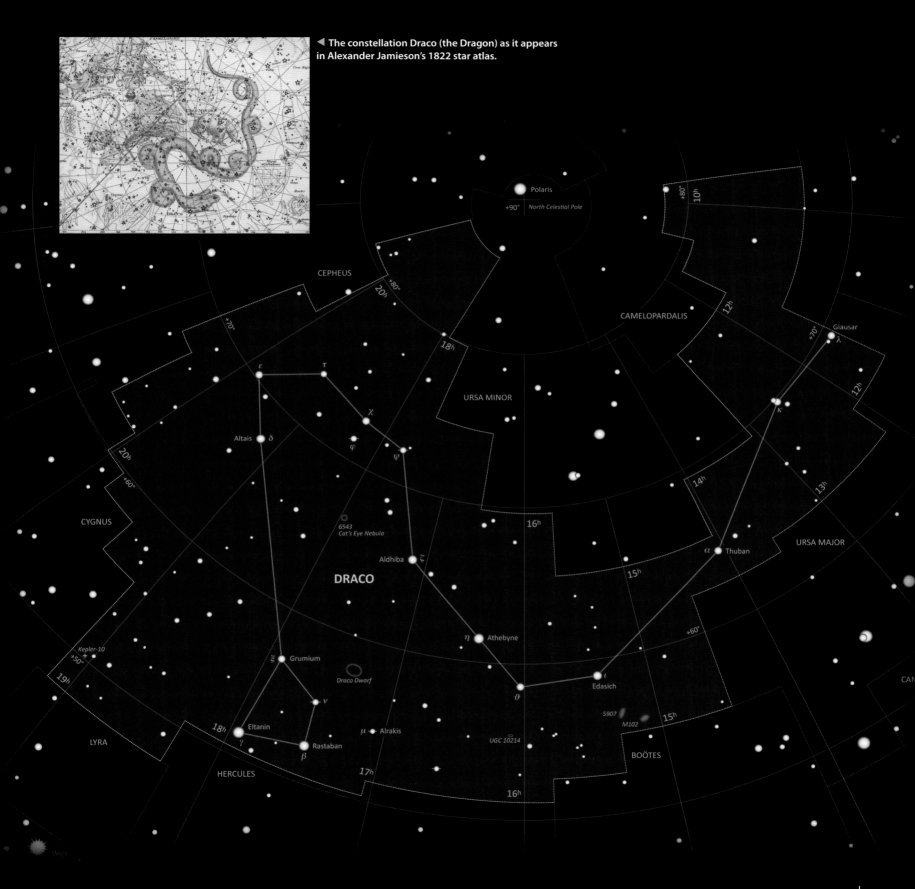

◀ **The constellation Draco (the Dragon) as it appears in Alexander Jamieson's 1822 star atlas.**

Polaris

+90° *North Celestial Pole*

CEPHEUS

URSA MINOR

CAMELOPARDALIS

Giausar
λ

κ

α Thuban

URSA MAJOR

+80°

ε τ

χ

φ ψ

Altais δ

6543
Cat's Eye Nebula

Aldhiba ζ

DRACO

η Athebyne

Kepler-10

Grumium ξ

Draco Dwarf

ι
Edasich

ϑ

ν

Eltanin Rastaban
γ β
Alrakis
μ

UGC 10214

5907

M102

CYGNUS

LYRA

HERCULES

BOÖTES

CAN

Vega

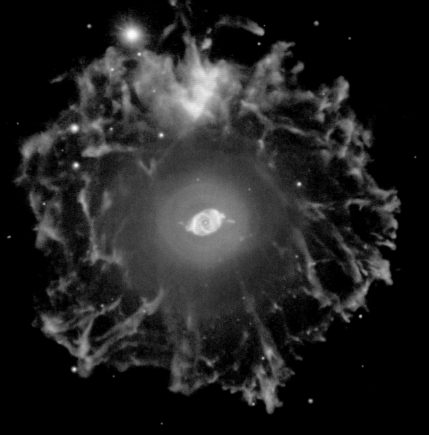

1725   By observing the star Gamma Draconis, or γ Dra, James Bradley and Samuel Molyneux discover the so-called aberration of starlight—the tiny periodic displacement of a star as a result of the continuous change in the direction of motion of Earth in its orbit around the sun. The effect is a result of the finite speed of light.

1781   French astronomer Pierre Méchain discovers the Spindle galaxy in Draco, also known as M102—a galaxy dissected by a central dust band that is seen exactly edge-on.

1786   William Herschel conducts the first observations of what is now known as the Cat's Eye Nebula (NGC 6543): a planetary nebula in Draco at some 3,000 light-years' distance. Its remarkable shape could be the result of the central star of the nebula being part of a binary.

1954   Studying photographic plates from the Palomar Observatory, Albert Wilson discovers the Draco Dwarf galaxy, one of the first dwarf companions of our Milky Way to be found. It is basically a flattened concentration of faint stars and huge amounts of dark matter, at a distance of some 260,000 light-years.

2011   The first terrestrial (Earth-like) exoplanet is discovered in Draco, around the star Kepler-10. But although Kepler-10b has a rocky composition, it orbits so close to its parent star that its surface is expected to be a huge sea of molten lava.

2017   In the galaxy NGC 5907 in Draco, at a distance of some 50 million light-years, astronomers find the most energetic pulsar ever. Every second, it produces as much energy as the sun does in 3.5 years.

▲ A wide-angle, long exposure photo of the Cat's Eye Nebula reveals that it is actually 3 light-years across.

◄ The Spindle galaxy, also known as M102 or NGC 5866, is seen almost exactly edge-on.

► The rocky planet Kepler-10b is so hot that it is covered by a planet-wide sea of lava.

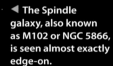

EQUULEUS

PASSPORT

Latin name: Equuleus	**Area:** 71.6 square degrees
English name: Foal	**Number of naked-eye stars:** 16
Genitive: Equulei	**Bordering constellations:** Delphinus, Aquarius, Pegasus
Abbreviation: Equ	**Best visibility:** July–September, north of 75° south
Origin: Ptolemy	

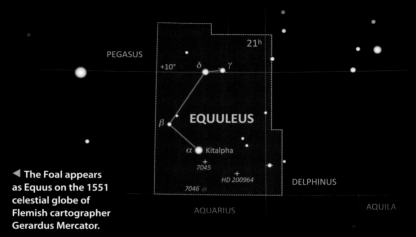

◀ The Foal appears as Equus on the 1551 celestial globe of Flemish cartographer Gerardus Mercator.

EQUULEUS (THE FOAL) is the second-smallest constellation in the sky. But unlike Crux (the Southern Cross), which is even smaller, Equuleus is very faint and hard to recognize. It can be found southeast of Delphinus (the Dolphin). In Greek mythology, it probably represents Hippe, daughter of the centaur Chiron and the nymph Chariclo.

▼ The giant extrasolar planet HD 200964 might have a moon that more or less resembles our own Earth.

TIMELINE

1790   William Herschel discovers the spiral galaxy NGC 7046 in Equuleus.

1827   His son, John Herschel, finds what he believes to be another galaxy in the small constellation: NGC 7045. It later turns out to be a binary star.

2010   Two Jupiter-like exoplanets are found orbiting the orange giant star HD 200964 in Equuleus, at 223 light-years' distance from Earth. The planets have orbital periods of 614 and 825 days, respectively.

▶ The barred spiral galaxy NGC 7046 is some 180 million light-years away.

ERIDANUS

P A S S P O R T

Latin name: Eridanus	**Area:** 1,137.9 square degrees
English name: River	**Number of naked-eye stars:** 194
Genitive: Eridani	**Bordering constellations:** Taurus, Cetus, Fornax, Phoenix, Tucana (corner), Hydrus, Horologium, Caelum, Lepus, Orion
Abbreviation: Eri	
Origin: Ptolemy	**Best visibility:** October–December, south of 30° north

TAURUS

ORION

4ʰ

3ʰ

0°

5ʰ

μ

ν

Beid

β Cursa

o¹

o²

Keid

I.2118
Witch Head Nebula

δ

ε Ran

η

Azha

Rigel

−10°

−10°

1535

LEPUS

Zaurak

γ

CETUS

ERIDANUS

1300

1232

−20°

−20°

τ⁴

MACS J0416.1-2403

FORNAX

CMB Cold Spot

COLUMBA

−30°

υ¹

υ²

CAELUM

4ʰ

−40°

ϑ

Acamar

−40°

1291

PICTOR

PHOENIX

HOROLOGIUM

3ʰ

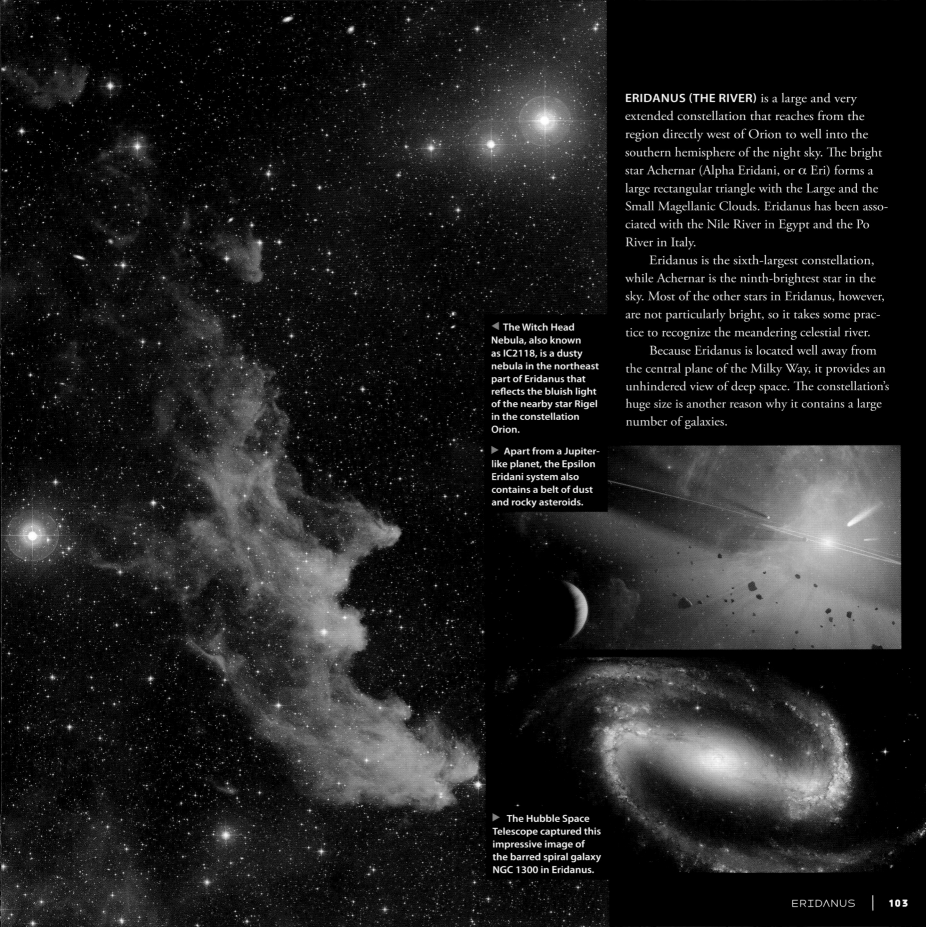

ERIDANUS (THE RIVER) is a large and very extended constellation that reaches from the region directly west of Orion to well into the southern hemisphere of the night sky. The bright star Achernar (Alpha Eridani, or α Eri) forms a large rectangular triangle with the Large and the Small Magellanic Clouds. Eridanus has been associated with the Nile River in Egypt and the Po River in Italy.

Eridanus is the sixth-largest constellation, while Achernar is the ninth-brightest star in the sky. Most of the other stars in Eridanus, however, are not particularly bright, so it takes some practice to recognize the meandering celestial river.

Because Eridanus is located well away from the central plane of the Milky Way, it provides an unhindered view of deep space. The constellation's huge size is another reason why it contains a large number of galaxies.

◀ The Witch Head Nebula, also known as IC2118, is a dusty nebula in the northeast part of Eridanus that reflects the bluish light of the nearby star Rigel in the constellation Orion.

▶ Apart from a Jupiter-like planet, the Epsilon Eridani system also contains a belt of dust and rocky asteroids.

▶ The Hubble Space Telescope captured this impressive image of the barred spiral galaxy NGC 1300 in Eridanus.

1784   English astronomer William Herschel discovers the galaxy NGC 1232, a beautiful spiral galaxy seen almost exactly face-on. Its distance is about 60 million light-years.

1835   William's son, John, observing from South Africa, hits upon another spectacular galaxy: NGC 1300, also at 60 million light-years. It is a picture-perfect example of a barred spiral galaxy, with an elongated central structure.

1960   The star Epsilon Eridani, or ε Eri, also known as Ran, is one of the targets for Frank Drake's Project Ozma (together with Tau Ceti, or τ Cet)—the first SETI program (Search for Extra-Terrestrial Intelligence). Epsilon Eridani is just 10.5 light-years away and quite similar to the sun.

2000   A Jupiter-like exoplanet, called Ægir, or ε Eri b, is discovered orbiting the star Epsilon Eridani once every seven years. The find is still somewhat controversial, but the existence of a dust belt around the star suggests that planets may indeed have formed. There's also indirect evidence for a second planet, ε Eri c.

2002   In *Star Trek Star Charts: The Complete Atlas of Star Trek*, Geoffrey Mandel states that Mr. Spock's home planet Vulcan orbits the star 40 Eridani A, which is also known as Omicron² Eridani (o² Eri), or Keid. The star is located in Sector 001 of the Alpha Quadrant of the *Star Trek* universe. In reality, 40 Eridani A is not known to have a planetary system.

◄ **Thanks to its rapid rotation, the bright star Achernar is extremely flattened at the poles.**

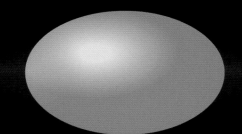

2003   Detailed measurements by a team of French astronomers using the European Very Large Telescope in Chile reveal that Achernar (Alpha Erdiani, or α Eri) is an extremely flattened star. As a result of its rapid rotation, its equatorial diameter is 50 percent larger than its polar diameter.

2004   In observations of the Wilkinson Microwave Anisotropy Probe, cosmologists discover a giant "cold spot" in the cosmic background radiation (CMB)—the "afterglow" of the big bang. The CMB Cold Spot happens to coincide with the Eridanus Supervoid—a galaxy-poor region in space, about 1 billion light-years across. The existence of such a large anisotropy in both the background radiation and the cosmic galaxy distribution is still poorly understood.

2014   Using the Hubble Space Telescope, astronomers complete their observations of the galaxy cluster MACS J0416.1-2403 in Eridanus, at a distance of 4 billion light-years from Earth. The observations are part of the Frontier Fields program. The massive cluster acts as a giant gravitational lens, enabling the detailed study of much more distant galaxies.

▶ **Blue colors denote the distribution of dark matter in the galaxy cluster MACS J0416.1-2403, as derived from the gravitational lensing effect on the shapes of background galaxies.**

The dark blue blob on the right side of this spherical map is the "cold spot" in the cosmic background radiation that coincides with the Eridanus Supervoid.

NGC 1232 is a beautiful face-on spiral galaxy with a small companion.

MACS J0416.1-2403

CMB Cold Spot

LEPUS

FORNAX

CAELUM

−30°

4ʰ

−40°

1291

ϑ

Acamar

−40°

HOROLOGIUM

ERIDANUS

PHOENIX

3ʰ

−50°

2ʰ

φ

χ

2ʰ

Achernar

α

HYDRUS

υ¹

υ²

FORNAX

PASSPORT

Latin name: Fornax	**Area:** 397.5 square degrees
English name: Oven	**Number of naked-eye stars:** 59
Genitive: Fornacis	**Bordering constellations:** Cetus, Sculptor, Phoenix, Eridanus
Abbreviation: For	**Best visibility:** October–November, south of 50° north
Origin: de Lacaille	

FORNAX (THE OVEN) is a rather obscure constellation in the southern sky, first listed as a separate constellation in the mid-eighteenth century. Even its brightest star, Alpha Fornacis, or α For, is difficult to see from populated areas as a result of light pollution. The constellation is home to the relatively close Fornax cluster of galaxies, but it also contains countless faint and remote galaxies, thanks to the fact that it lies far away from the obscuring band of our own Milky Way.

◀ The Fornax cluster of galaxies contains some sixty members, at an average distance of about 60 million light-years. NGC 1365 is in the lower right.

▶ The European Very Large Telescope made this photo of the barred spiral galaxy NGC 1097 and its small companion (upper right).

▲ Long exposures of the Hubble Ultra Deep Field have revealed many thousands of galaxies at the edge of the observable universe.

▲ The Fornax Dwarf galaxy was one of the first dwarf companions of the Milky Way to be recognized as such.

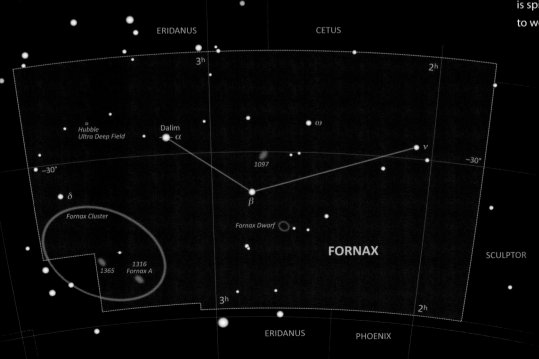

◄ Orange colors denote the radio emission of Fornax A, the fourth-brightest radio source in the sky. The radio waves are produced by electrons blown into space by the central elliptical galaxy NGC 1316.

TIMELINE

1763   French astronomer Nicolas Louis de Lacaille introduces the constellation as Fornax Chemica (the Chemical Oven); the name is later shortened to Fornax.

1790   English astronomer William Herschel discovers the beautiful barred spiral galaxy NGC 1097, just north of the star Beta Fornacis, or β For. The active galaxy is about 45 million light-years away.

1938   American astronomer Harlow Shapley is the first to note a concentration of faint stars in Fornax, now known as the Fornax Dwarf galaxy. It is one of the dozens of dwarf companions of our own Milky Way, at a distance of some 460,000 light-years. The dwarf galaxy contains six globular clusters, one of which had already been found before Shapley made his discovery.

2003   The Hubble Space Telescope carries out the first observations of the Hubble Ultra Deep Field (HUDF)—a small region of sky in Fornax without any foreground objects. Eventually, long observations at ultraviolet, visible, and infrared wavelengths reveal more than ten thousand extremely remote galaxies in the HUDF, providing astronomers with a look into the earliest history of the universe.

2013   Observations by NASA's NuSTAR X-ray satellite reveal that the supermassive black hole at the heart of the barred spiral galaxy NGC 1365 is spinning at almost the speed of light. The black hole is estimated to weigh about 2 million times as much as the sun.

▲ Hidden in the core of the barred spiral galaxy NGC 1365 is a supermassive black hole that spins around almost at the speed of light.

GEMINI

PASSPORT

Latin name: Gemini		**Area:** 513.8 square degrees	
English name: Twins		**Number of naked-eye stars:** 119	
Genitive: Geminorum		**Bordering constellations:** Auriga, Taurus, Orion, Monoceros, Canis Minor, Cancer, Lynx	
Abbreviation: Gem			
Origin: Ptolemy		**Best visibility:** December–January, north of 50° south	

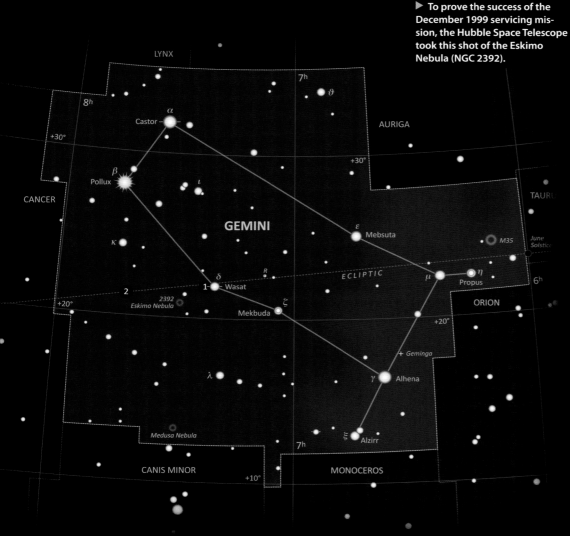

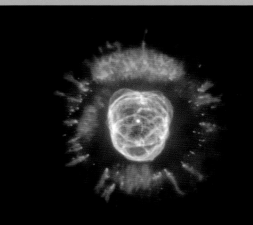

▶ To prove the success of the December 1999 servicing mission, the Hubble Space Telescope took this shot of the Eskimo Nebula (NGC 2392).

▶ The surface of the large Saturnian moon Titan, as imaged by the European Huygens probe after its descent in January 2005.

GEMINI (THE TWINS) is one of the most prominent constellations of the zodiac. In northern winter months, it's easy to spot, thanks to its two brightest stars, Castor and Pollux. The sun passes through the constellation between June 21 and July 20, during the height of the northern summer.

In Greek mythology, Castor was the mortal son of King Tyndareus of Sparta and his wife Leda, while Pollux (also known as Polydeuces) was conceived by Zeus. When Castor died in a battle, Pollux asked his father to be reunited with his brother again, after which Zeus placed both of them in the heavens.

The Milky Way passes through the western part of Gemini, and the constellation contains quite a number of star clusters and nebulae. It also served as the backdrop for the discovery of Pluto and the descent of the European Huygens probe to the surface of the Saturnian moon Titan.

▶ **The Medusa Nebula (also known as Abell 21 or Sh 2-274) is a planetary nebula consisting of filamentary wisps of gas.**

▼ **Discovery photos of Pluto (arrowed), a tiny speck of light, slowly moving through the constellation Gemini in early 1930.**

1745 Swiss astronomer Jean-Philippe de Chéseaux is the first one to spot the open star cluster M35, at the western edge of Gemini. The cluster, at a distance of 2,800 light-years, contains some 2,500 stars and is easily visible in binoculars.

1787 Scanning the constellation Gemini with his homemade telescopes, English astronomer William Herschel discovers a planetary nebula that is now known as the Eskimo Nebula, or the Clown Face Nebula (NGC 2392), because of its resemblance to a human face.

1930 ❶ On February 18, American astronomer Clyde Tombaugh discovers the dwarf planet Pluto, on photographic plates made a few weeks earlier at Lowell Observatory in Flagstaff, Arizona. At that time, Pluto is close in the sky to the star Wasat (Delta Geminorum, or δ Gem). For seventy-six years, Pluto was listed as the ninth planet in our solar system.

1955 George Abell, known for his work on galaxy clusters, discovers the Medusa Nebula in Gemini. Initially, it was thought to be the remnant of a supernova explosion, but astronomers now realize it is a planetary nebula at a distance of some 1,500 light-years.

1975 NASA's Second Small Astronomy Satellite picks up energetic gamma rays from a source in Gemini—the first unidentified cosmic gamma-ray source to be found. Called Geminga, it is now known to be an isolated neutron star at a mere 800 light-years away.

2005 ❷ On January 14, NASA's Deep Space Network is aimed at Gemini to pick up signals from the European Huygens probe during its parachuted descent to the bitterly cold and methane-rich surface of Titan, the largest moon of Saturn. Meanwhile, the American Cassini spacecraft is orbiting the planet, which is located due south of Castor during the Huygens mission.

2006 A periodic wobble in the radial velocity (toward and away from the Earth) of Pollux (Beta Geminorum, or β Gem) reveals the existence of a rather massive Jupiter-like planet (Pollux b), orbiting the giant star once every 590 days. In 2015, the planet is officially named Thestias.

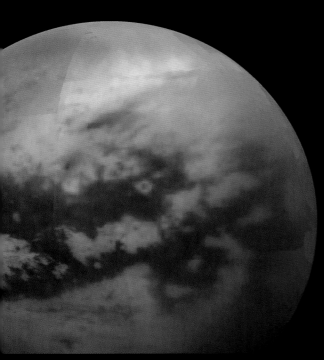

◀ When Saturn was in Gemini in early 2005, the Huygens probe descended to the surface of the giant moon Titan, which is seen here in an infrared image by NASA's planetary explorer Cassini.

◀ The open star cluster M35 in Gemini is some 25 light-years across and contains about 200 relatively young stars.

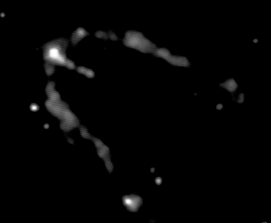

▲ This X-ray image reveals that Geminga produces two jets of particles that are swept backward as the neutron star races through the Milky Way galaxy.

GRUS

PASSPORT

Latin name: Grus	**Area:** 365.5 square degrees
English name: Crane	**Number of naked-eye stars:** 55
Genitive: Gruis	**Bordering constellations:** Piscis Austrinus, Microscopium, Indus, Tucana, Phoenix, Sculptor
Abbreviation: Gru	
Origin: Keyser and de Houtman	**Best visibility:** August–September, south of 30° north

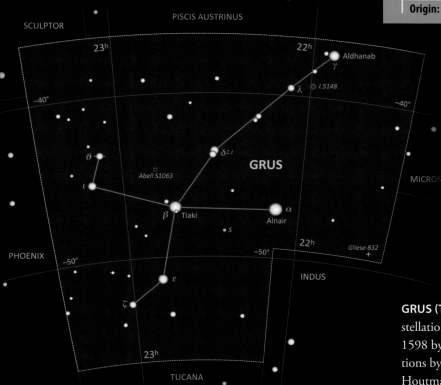

▲ The Spare Tire Nebula (IC 5148) is some 3,000 light-years away.

GRUS (THE CRANE) is a rather nondescript constellation in the southern sky, first introduced in 1598 by Petrus Plancius, on the basis of observations by Pieter Dirkszoon Keyser and Frederick de Houtman. Located far away from the central band of the Milky Way, it contains many remote galaxies.

▼ Dominated by a massive elliptical galaxy, the galaxy cluster Abell S1063 in Grus acts as a powerful gravitational lens.

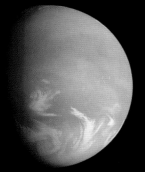

▲ Artist's impression of the potentially habitable super-Earth Gliese 832c.

TIMELINE

1894   Australian amateur astronomer Walter Gale discovers the Spare Tire Nebula (IC 5148), a planetary nebula in Grus with an extremely large expansion velocity of 30 miles per second.

2014   Orbiting at the edge of the habitable zone of the red dwarf star Gliese 832 in Grus, at 16 light-years' distance from the sun, a super-Earth is found (Gliese 832c)—one of the closest potentially habitable exoplanets known.

2016   The Hubble Space Telescope completes its Frontier Fields observations of the galaxy cluster Abell S1063, which acts as a gravitational lens, magnifying and distorting the images of remote background galaxies.

HERCULES

PASSPORT

Latin name: Hercules	**Area:** 1,225.1 square degrees
English name: Hercules	**Number of naked-eye stars:** 245
Genitive: Herculis	**Bordering constellations:** Draco, Boötes, Corona Borealis, Serpens Caput, Ophiuchus, Aquila, Sagitta, Vulpecula, Lyra
Abbreviation: Her	
Origin: Ptolemy	**Best visibility:** May–June, north of 35° south

▲ The Very Large Array radio telescope reveals the huge jets and "lobes" of the central elliptical galaxy 3C348, which harbors a supermassive black hole.

HERCULES IS A LARGE CONSTELLATION in the northern sky, but despite its size, it contains relatively few bright stars. Hercules is the Roman name of the Greek mythological hero Herakles. For most northern hemisphere observers, he appears to be standing upside down, with his head marked by the bright star Rasalgethi (Alpha Herculis, or α Her), close to the head of Ophiuchus (the Serpent Bearer). Hercules's foot is firmly placed on the head of Draco (the Dragon)—the monster he slayed as one of his twelve labors.

The most prominent sight in Hercules is the large globular cluster M13. On clear, moonless summer nights, it can just be seen by the naked eye, on the west side of the central "keystone" asterism—the trapezium-shaped quadrangle formed by the stars Eta (η), Zeta (ζ), Epsilon (ε), and Pi (π) Herculis.

◄ Hundreds of thousands of stars swarm around in the core of the globular cluster M13.

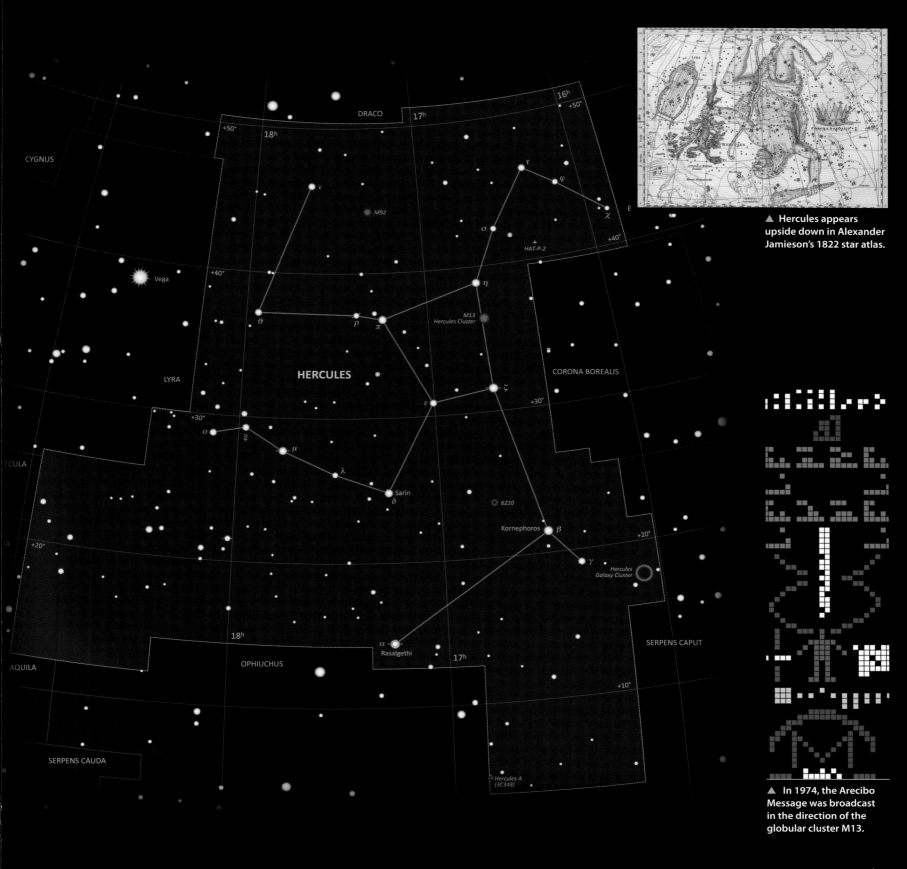

CYGNUS

DRACO

18ʰ

17ʰ

16ʰ

+50°

+50°

+50°

τ

φ

χ

E

ι

M92

σ

+40°

HAT-P-2

+40°

Vega

+40°

η

ϑ

ρ

π

M13
Hercules Cluster

CORONA BOREALIS

LYRA

HERCULES

ζ

+30°

ε

+30°

o

ꞯ

μ

PECULA

λ

Sarin

δ

6210

+20°

Kornephoros

β

+20°

γ

Hercules
Galaxy Cluster

ꞮA

+20°

18ʰ

α

Rasalgethi

17ʰ

SERPENS CAPUT

AQUILA

OPHIUCHUS

+10°

SERPENS CAUDA

Hercules A
(3C348)

▲ Hercules appears
upside down in Alexander
Jamieson's 1822 star atlas.

▲ In 1974, the Arecibo
Message was broadcast
in the direction of the
globular cluster M13.

TIMELINE

1714 English astronomer Edmund Halley is the first one to note a spherical nebulosity in Hercules, now known as M13. The globular cluster contains hundreds of thousands of stars in a region 150 light-years across. Its distance is approximately 22,000 light-years.

1777 In Germany, Johann Bode discovers another globular cluster in Hercules, M92, which is smaller, farther away, and dimmer than M13.

1783 William Herschel studies the motions of stars through the Milky Way galaxy and concludes that our sun is moving in the direction of the star Lambda Herculis, or λ Her. Herschel was almost right; astronomers now know that this so-called solar apex lies at the border between the constellations Hercules and Lyra (the Lyre).

1959 A source of radio waves in Hercules (Hercules A) is found to be associated with a remote elliptical galaxy, 3C348, at 2.1 billion light-years away. The radio waves are produced by huge, powerful jets of high-energy particles. The jets, in turn, are blown into space by a supermassive black hole in the galaxy's core, weighing in at some 2.5 billion times the mass of the sun.

1974 On the occasion of the inauguration of the 1,000-foot Arecibo radio telescope in Puerto Rico, a message containing information about humankind is sent in the direction of globular cluster M13. It is the first Messaging to Extraterrestrial Intelligence (METI) project ever.

2007 A very massive planet, HAT-P-2b, is found in a 5.6-day orbit around a star in Hercules, 370 light-years away. At 9 Jupiter masses, it was the most massive exoplanet known at the time of its discovery.

◀ At about half a billion light-years away is the Hercules cluster of galaxies, containing a few hundred individual galaxies.

HOROLOGIUM

PASSPORT

Latin name: Horologium	**Area:** 248.9 square degrees
English name: Clock	**Number of naked-eye stars:** 30
Genitive: Horologii	**Bordering constellations:** Eridanus, Hydrus, Reticulum, Dorado, Caelum
Abbreviation: Hor	
Origin: de Lacaille	**Best visibility:** October–December, south of 20° north

◀ **At a distance of 400,000 light-years, Arp-Madore 1 is the most remote globular cluster in our Milky Way galaxy.**

HOROLOGIUM (THE CLOCK) is an inconspicuous galaxy in the southern sky. Its southernmost part lies between the bright star Achernar and the Large Magellanic Cloud.

▶ **NGC 1448 is a spiral galaxy in Horologium that is seen almost exactly edge-on.**

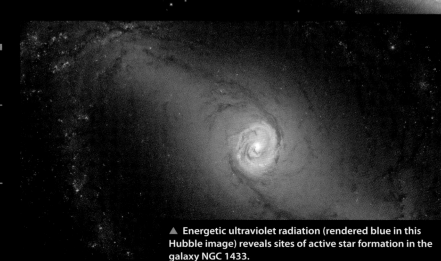

TIMELINE

1763 In his southern star catalog, Nicolas Louis de Lacaille introduces Horologium as a new constellation.

1826 James Dunlop discovers a number of star clusters and galaxies in Horologium, including NGC 1433. This barred spiral galaxy has an active nucleus and a remarkable double ring of newly formed stars.

1979 Astronomers Halton Arp and Barry Madore identify a small star cluster in Horologium as one of the farthest-known globular clusters in the Milky Way galaxy: Arp-Madore 1 is 400,000 light-years away.

▲ **Energetic ultraviolet radiation (rendered blue in this Hubble image) reveals sites of active star formation in the galaxy NGC 1433.**

HYDRA

PASSPORT

Latin name: Hydra		**Area:** 1,302.8 square degrees	
English name: Sea Serpent		**Number of naked-eye stars:** 238	
Genitive: Hydrae		**Bordering constellations:** Cancer, Canis Minor, Monoceros, Puppis, Pyxis, Antlia, Centaurus, Lupus (corner), Libra, Virgo, Corvus, Crater, Sextans, Leo	
Abbreviation: Hya			
Origin: Ptolemy			
		Best visibility: February–March, between 50° north and 80° south	

HYDRA (THE SEA SERPENT) is the largest constellation in the sky. Its head—a compact group of six stars—lies more or less between the bright stars Procyon and Regulus, while the tail of the serpentine creature is to be found south of Spica. The bright star Alphard (Alpha Hydrae, or α Hya) was known to sixteenth-century Danish astronomer Tycho Brahe as Cor Hydrae, the heart of the water snake.

Hydra has been associated with the nine-headed water snake of Lerna, which was slain by Hercules in the second of his twelve labors. In another mythological story, the creature is linked to the small constellations Corvus (the Raven) and Crater (the Cup), which are located on the snake's back.

Partly due to its huge size, Hydra contains a lot of interesting objects, including nearby baby stars and distant galaxies. It is also home to the most intensely observed astronomical event in history: the collision of two neutron stars, which produced a burst of gravitational waves.

◀ The Southern Pinwheel galaxy (M83) is one of the most photogenic galaxies in the southern sky.

◄ Dark gaps in the protoplanetary disk surrounding the young star TW Hydrae suggest that planets have already started to form.

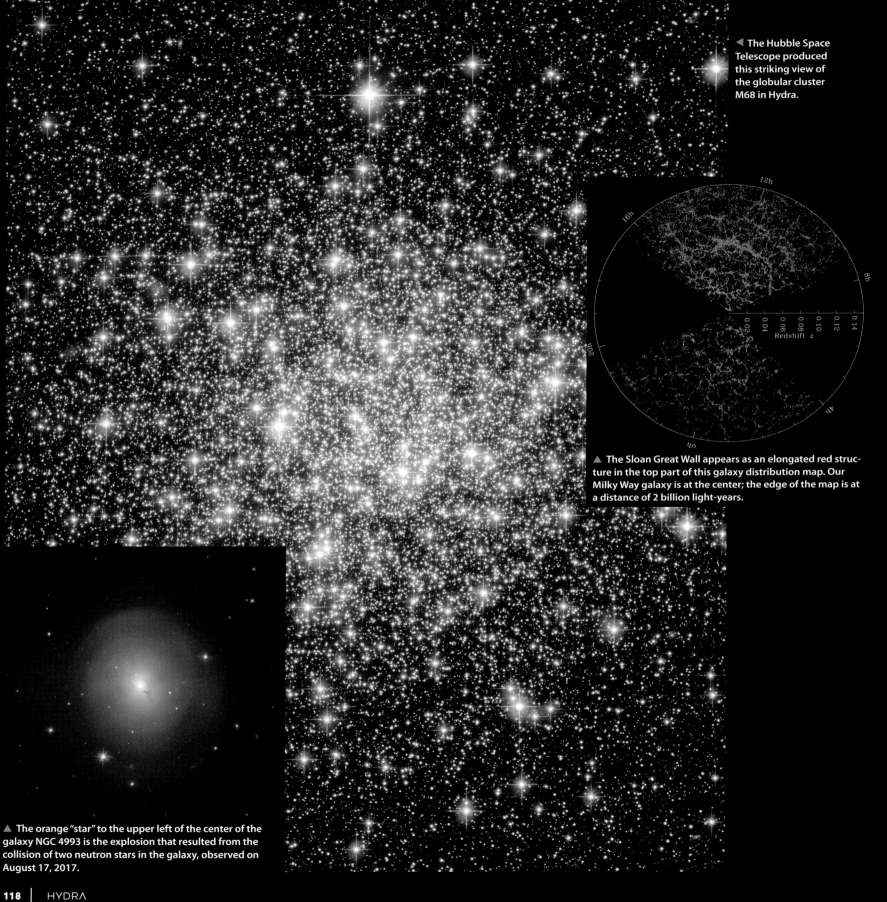

◄ The Hubble Space Telescope produced this striking view of the globular cluster M68 in Hydra.

▲ The Sloan Great Wall appears as an elongated red structure in the top part of this galaxy distribution map. Our Milky Way galaxy is at the center; the edge of the map is at a distance of 2 billion light-years.

▲ The orange "star" to the upper left of the center of the galaxy NGC 4993 is the explosion that resulted from the collision of two neutron stars in the galaxy, observed on August 17, 2017.

◀ Artist's impression of the collision and merger of two neutron stars that produced a burst of gravitational waves.

▶ M48 is a loose cluster of a few hundred stars, first listed by French astronomer Charles Messier.

TIMELINE

1752 Studying the southern sky from Cape Town, French astronomer Nicolas Louis de Lacaille discovers the galaxy M83, at the border of Hydra and Centaurus (the Centaur). Also known as the Southern Pinwheel galaxy, it is one of the nearest barred spiral galaxies known.

1771 In the northwestern part of the constellation, Charles Messier discovers a loose open cluster of stars. M48, as it is called, is 300 million years old and lies at a distance of 1,500 light-years.

1780 Messier is also the discoverer of M68. This globular cluster is some 33,000 light-years away. It lies just south of the star Beta Corvi, or β Cor.

2003 Observations of the Sloan Digital Sky Survey reveal the existence of the Sloan Great Wall, a huge supercluster of galaxies, about 1 billion light-years away. It has a length of 1.4 billion light-years. As seen from Earth, most of its galaxies lie within the borders of the constellation Hydra. It ranks as one of the largest structures in the observable universe.

2016 The ALMA observatory in northern Chile (using the Atacama Large Millimeter/submillimeter Array) produces the most detailed face-on image of the dusty protoplanetary disk around a newborn star. TW Hydrae, at 175 light-years from Earth, is just 10 million years old. Concentric gaps in the dust disk suggest that planets have already started to form.

2017 On August 17, gravitational-wave observatories in the United States and Europe detect faint ripples in the fabric of space-time. The gravitational waves coincide with an explosion in the galaxy NGC 4993, 130 million light-years away in Hydra. The event is studied by dozens of observatories, both on the ground and in space. The observations indicate that two compact neutron stars have collided and merged into a black hole. In the process, huge amounts of heavy elements, including gold and platinum, have been synthesized and blown into space.

HYDRUS

PASSPORT

Latin name: Hydrus	**Area:** 243.0 square degrees
English name: Water Snake	**Number of naked-eye stars:** 33
Genitive: Hydri	**Bordering constellations:** Eridanus, Phoenix (corner), Tucana, Octans, Mensa, Dorado, Reticulum, Horologium
Abbreviation: Hyi	
Origin: Keyser and de Houtman	**Best visibility:** September–November, south of 5° north

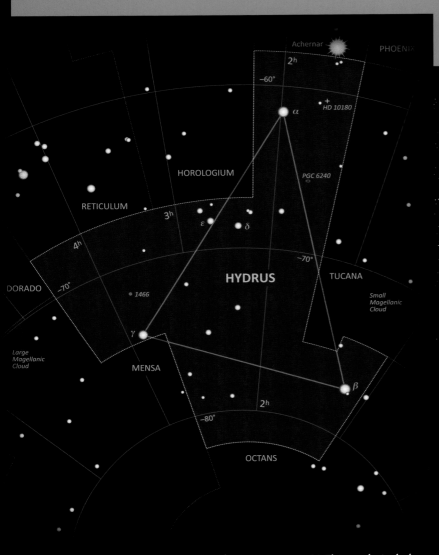

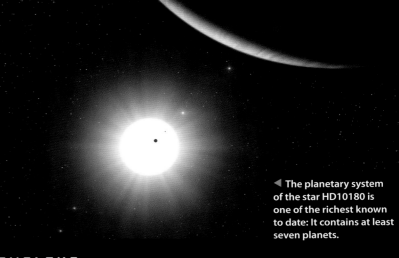

◀ **The planetary system of the star HD10180 is one of the richest known to date: It contains at least seven planets.**

HYDRUS (THE WATER SNAKE) is a relatively large constellation in the southern sky. Its three brightest stars form a triangle that lies between the bright star Achernar and the south celestial pole.

◀ **NGC 1466 is a relatively small globular cluster in Hydrus, containing a large number of variable stars.**

TIMELINE

1598   The constellation Hydrus is introduced by Petrus Plancius, on the basis of observations by Dutch sailors Pieter Dirkszoon Keyser and Frederick de Houtman.

1834   From his observatory near Cape Town, South Africa, John Herschel discovers the globular cluster NGC 1466 in Hydrus, at a distance of 160,000 light-years.

2010   Tiny periodic variations in the radial velocity (toward or away from us) of the star HD 10180, 127 light-years away in Hydrus, reveal the existence of a system of at least seven and possibly even nine planets. It is one of the richest planetary systems known to date.

▶ **Nicknamed the White Rose galaxy, PGC 6240 is a galaxy in Hydrus surrounded by concentric shells of gas.**

INDUS

INDUS (THE INDIAN) is a faint southern constellation conceived by Dutch sailors Pieter Dirkszoon Keyser and Frederick de Houtman to represent one of the indigenous people they encountered on their trips.

PASSPORT

Latin name: Indus	**Area:** 294.0 square degrees
English name: Indian	**Number of naked-eye stars:** 42
Genitive: Indi	**Bordering constellations:** Microscopium, Sagittarius (corner), Telescopium, Pavo, Octans, Tucana, Grus
Abbreviation: Ind	
Origin: Keyser and de Houtman	**Best visibility:** July–September, south of 15° north

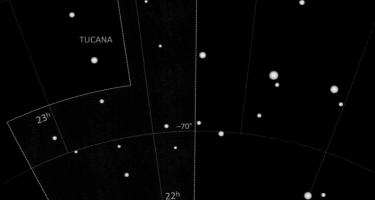

▶ Two brown dwarfs (foreground) orbit the star Epsilon Indi (left). Our own sun is in the background (upper right).

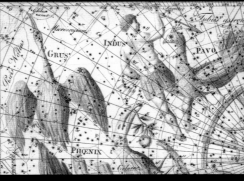

▲ Armed with arrows, Indus features in Johann Bode's *Uranographia*.

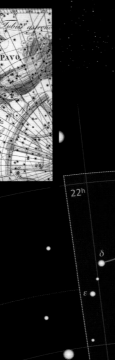

◀ A ring of dark dust is silhouetted against the bright glow of the elliptical galaxy NGC 7049 in Indus.

TIMELINE

1960 The nearby star Epsilon Indi (or ε Ind) is one of the targets of Frank Drake's Project Ozma, the first SETI (Search for Extraterrestrial Intelligence) program.

2003 A binary brown dwarf is discovered in a very wide orbit around the star Epsilon Indi. At the time, those were the nearest brown dwarfs known.

2018 Astronomers confirm the existence of a Jupiter-like planet orbiting Epsilon Indi. At just under 12 light-years, it is the nearest known gas giant outside our solar system.

N 1717, BRITISH ASTRONOMER EDMUND HALLEY (most famous for the comet named after him) discovered something peculiar. Two bright stars, Arcturus in the constellation Boötes and Sirius in Canis Major, appeared to be at slightly different positions in the sky than they had been many centuries ago, as recorded in Claudius Ptolemy's *Almagest*. Halley's rightful conclusion: Stars are not at fixed positions. Instead, they have their own motion through space.

We now know that the stars in the night sky actually orbit the center of our Milky Way galaxy at velocities of a few hundred miles per second. Our own sun is no exception. This galactic rotation, however, isn't perfectly smooth. In other words, stars also move about relative to one another. And unless a star happens to move exactly toward us or away from us, this spatial motion results in a positional change on the sky.

A low-flying bird crosses your field of vision much faster than an airplane high up in the sky, even though the velocity of the plane is much higher. For the same reason, stars that are relatively close, like Arcturus and Sirius, display a larger proper motion across the sky than stars that are farther away. Barnard's Star, a faint red dwarf star in the constellation Ophiuchus at a distance of just 6 light-years, has the largest proper motion of any star: 10.3 arc seconds per year. This corresponds to the apparent diameter of the moon in just 175 years.

While the proper motion of stars is much too slow to be evident during our lifetime, it nevertheless changes the appearance of the familiar constellations over time, as is shown in the illustration on this page. About a hundred thousand years ago, when our distant ancestors were moving out of Africa and starting to spread around the globe, the Big Dipper (the seven brightest stars of the constellation Ursa Major) looked more like a fly swatter. Another hundred thousand years in the future, ye seven stars will resemble a swimming duck

Because of stellar proper motion, our constellations are just temporary patterns in the night sky. In fact, their appearance is changing more rapidly than the appearance of our home planet. Just like the constellations, Earth's continents also slowly change their shape and relative positions, but continental drift plays itself out at a much slower pace. If you could travel back in time 1 or 2 million years, North America, Europe, and Australia would look more or less the same, but the starry sky would be completely unfamiliar.

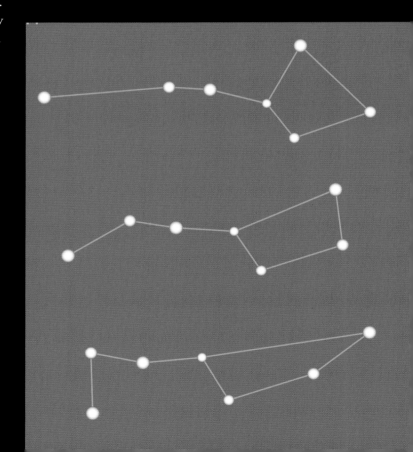

The Big Dipper
Top: 100,000 BC
Middle: Today
Bottom: 100,000 AD

LACERTA

PASSPORT

Latin name: Lacerta	**Area:** 200.7 square degrees
English name: Lizard	**Number of naked-eye stars:** 68
Genitive: Lacertae	**Bordering constellations:** Cepheus, Cygnus, Pegasus, Andromeda, Cassiopeia
Abbreviation: Lac	
Origin: Hevelius	**Best visibility:** August–September, north of 30° south

LACERTA (THE LIZARD) is a small constellation in the northern sky, between Cassiopeia and Cygnus (the Swan), with a characteristic zigzag shape. Its northern half lies in the band of the Milky Way.

▶ **BL Lacertae is the prototype of a certain class of black hole–powered active galaxies, spewing energetic jets into space.**

TIMELINE

1687 Lacerta is introduced as a new constellation by Polish astronomer Johannes Hevelius, in his star atlas *Firmamentum Sobiescianum*.

1968 BL Lacertae, listed as a variable star in 1929, turns out to be a distant galaxy with a highly variable, active nucleus. It has become the prototype of the BL Lac class of active galaxies, which are powered by supermassive black holes.

2008 The red dwarf star EV Lacertae, at just 16 light-years' distance, produces the brightest stellar flare ever observed, thousands of times more energetic than the most powerful solar flares.

▲ A foreground star in our own Milky Way galaxy almost outshines the glow of galaxy NGC 7250, 45 million light-years away in Lacerta.

◀ Artist's impression of the extremely powerful flare produced by the red dwarf star EV Lacertae in April 2008.

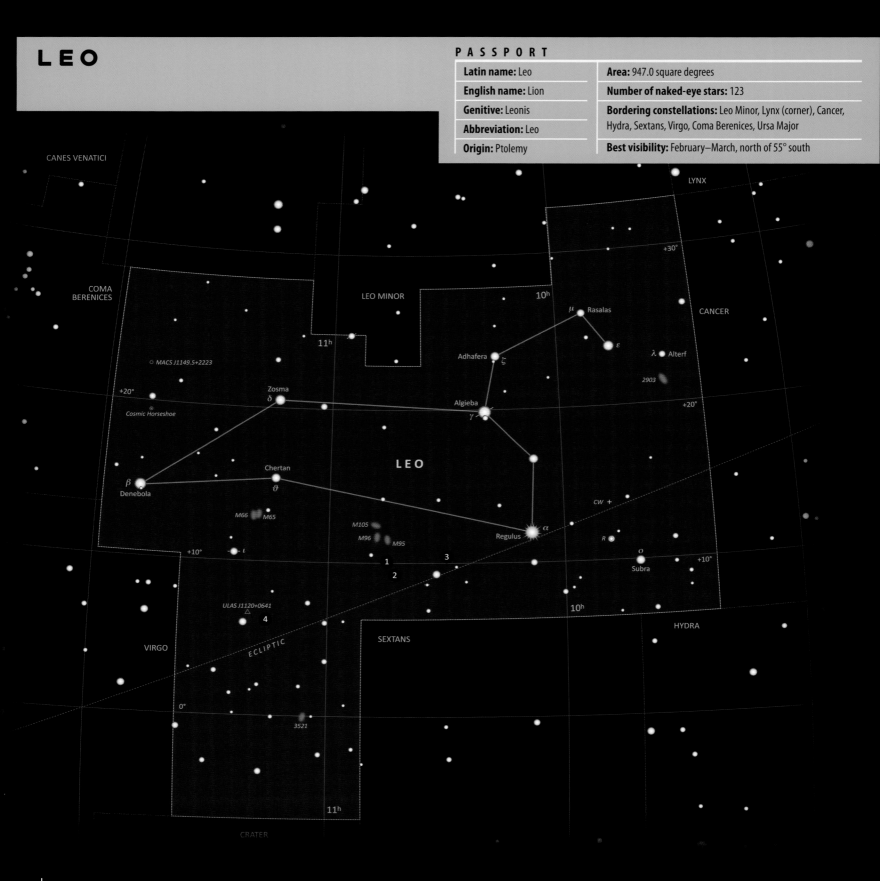

LEO

PASSPORT

Latin name: Leo	**Area:** 947.0 square degrees
English name: Lion	**Number of naked-eye stars:** 123
Genitive: Leonis	**Bordering constellations:** Leo Minor, Lynx (corner), Cancer, Hydra, Sextans, Virgo, Coma Berenices, Ursa Major
Abbreviation: Leo	
Origin: Ptolemy	**Best visibility:** February–March, north of 55° south

COMA
BERENICES

CANCER

LEO MINOR

10ʰ

μ Rasalas

11ʰ

Adhafera
ζ

ε

λ Alterf

2903

MACS J1149.5+2223

+20°

Zosma
δ

Algieba
γ

+20°

Cosmic Horseshoe

LEO

Chertan

CW +

β
Denebola

ϑ

M66 M65

M105

M96 M95

Regulus α

R

o
Subra

+10°
ι

1
2

3

+10°

ULAS J1120+0641
4

10ʰ

HYDRA

VIRGO

ECLIPTIC

SEXTANS

0°

3521

11ʰ

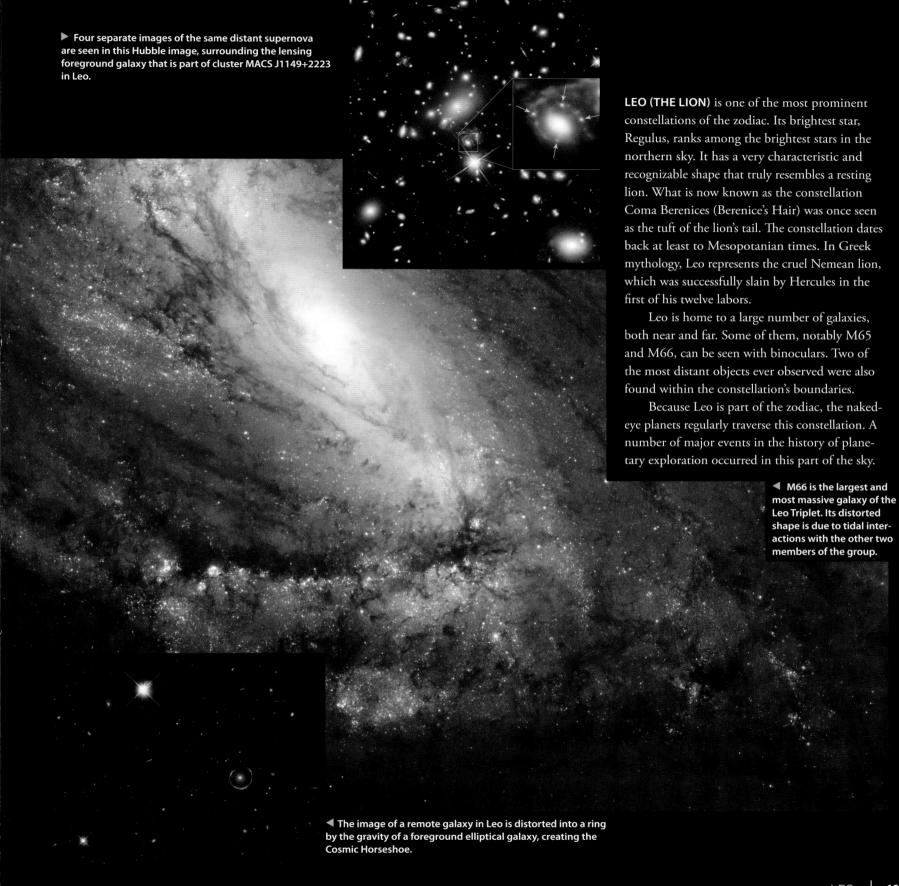

▶ Four separate images of the same distant supernova are seen in this Hubble image, surrounding the lensing foreground galaxy that is part of cluster MACS J1149+2223 in Leo.

LEO (THE LION) is one of the most prominent constellations of the zodiac. Its brightest star, Regulus, ranks among the brightest stars in the northern sky. It has a very characteristic and recognizable shape that truly resembles a resting lion. What is now known as the constellation Coma Berenices (Berenice's Hair) was once seen as the tuft of the lion's tail. The constellation dates back at least to Mesopotanian times. In Greek mythology, Leo represents the cruel Nemean lion, which was successfully slain by Hercules in the first of his twelve labors.

Leo is home to a large number of galaxies, both near and far. Some of them, notably M65 and M66, can be seen with binoculars. Two of the most distant objects ever observed were also found within the constellation's boundaries.

Because Leo is part of the zodiac, the naked-eye planets regularly traverse this constellation. A number of major events in the history of planetary exploration occurred in this part of the sky.

◀ M66 is the largest and most massive galaxy of the Leo Triplet. Its distorted shape is due to tidal interactions with the other two members of the group.

◀ The image of a remote galaxy in Leo is distorted into a ring by the gravity of a foreground elliptical galaxy, creating the Cosmic Horseshoe.

▶ The carbon star CW Leonis has blown a shell of gas and dust into space almost 3 light-years across.

TIMELINE

1655 ❶ On March 25, Dutch astronomer Christiaan Huygens observes Saturn when it is high in the sky in Leo, and discovers its largest moon, Titan. Titan is even larger than the planet Mercury. It's also the only planetary moon with a substantial atmosphere.

1780 French astronomer Charles Messier discovers two spiral galaxies in Leo, M65 and M66. Together with the nearby galaxy NGC 3628, they form the so-called Leo Triplet galaxy group, at just under 40 million light-years' distance.

1969 The variable star CW Leonis is found to be a so-called carbon star—actually the nearest to Earth, at some 400 light-years. It blows shells of carbon-rich dust into space, while it speeds through the Milky Way galaxy at 55 miles per second.

1976 ❷ On July 4, NASA's Viking 1 spacecraft completes the first truly successful soft landing on the surface of Mars, in an area known as Chryse Planitia. At the time, Mars is slightly east of the star Rho Leonis, or ρ Leo. Viking 1 is equipped with instruments to search for any form of biological activity on the planet, but the results remain inconclusive.

2003 ❸ Jupiter is passing through the constellation Leo when the Galileo orbiter makes its lethal plunge into the giant planet's thick atmosphere, ending its eight-year mission of exploration. Galileo was the first spacecraft to orbit a giant planet.

2007 Astronomers find a textbook example of a gravitational lens in Leo. Known as the Cosmic Horseshoe, it is an almost complete Einstein ring—the image of a background galaxy hugely distorted by the gravity of a massive foreground object.

2011 The United Kingdom Infra-Red Telescope at Mauna Kea, Hawaii, is used to discover the second most distant quasar, in the southern part of Leo. Known as ULAS J1120+0641, it is the luminous core of a young, remote galaxy harboring a supermassive black hole.

2014 For the first time ever, astronomers observe multiple and successive images of the same remote supernova explosion. Thanks to the gravitational lensing effect of a foreground cluster of galaxies in Leo (MACS J1149.5+2223, one of the targets of the Frontier Fields program of the Hubble Space Telescope), the supernova's light arrives at Earth along a number of different paths.

2016 ❹ While Jupiter passes through the southern part of Leo, NASA's Juno spacecraft is inserted in an elliptical orbit around the giant planet on July 4. Juno is the first spacecraft to fly over Jupiter's north and south poles, revealing a turbulent world of cyclones.

▶ The first-ever image taken from the surface of Mars was obtained by the Viking 1 lander in 1976.

▲ Saturn's giant moon Titan (background) was discovered in 1655 by Christiaan Huygens. Dione (foreground) was found in 1684.

▶ Eight smaller cyclones surround the single larger one at Jupiter's north pole, as seen by NASA's Juno spacecraft.

LEO MINOR

PASSPORT

Latin name: Leo Minor

English name: Lesser Lion

Genitive: Leonis Minoris

Abbreviation: LMi

Origin: Hevelius

Area: 232.0 square degrees

Number of naked-eye stars: 37

Bordering constellations: Ursa Major, Lynx, Cancer (corner), Leo

Best visibility: February–March, north of 45° south

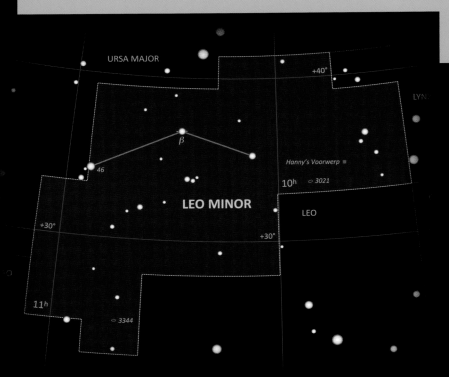

LEO MINOR (THE LESSER LION) is a small and relatively new constellation, riding on the back of Leo (the Lion). It contains few bright stars, but many galaxies. The most famous object in Leo Minor is the mysterious Hanny's Voorwerp.

◀ **Seen almost face-on, NGC 3344 is a beautiful spiral galaxy at 20 million light-years away in Leo Minor.**

TIMELINE

1687   Polish astronomer Johannes Hevelius introduces Leo Minor as one of his ten new constellations—a tiny grouping beneath one of the paws of Ursa Major (the Great Bear).

1995   A supernova explodes in NGC 3021, a face-on spiral galaxy at some 100 million light-years' distance in Leo Minor. Observations of the exploding star are used to further calibrate the distance scale of the universe.

2007   Dutch schoolteacher Hanny van Arkel, taking part in the citizen science project Galaxy Zoo, discovers a mysterious, greenish Milky Way–sized blob of light close to the remote galaxy IC 2497, 650 million light-years away in Leo Minor. Hanny's Voorwerp, as it is called, is probably an intergalactic cloud of gas, energized by former quasar-like activity in IC 2497.

▲ **NGC 3021 is a galaxy in Leo Minor in which a supernova explosion was observed in 1995.**

▶ **A green blob of interstellar gas, close to the galaxy IC 2497, is named Hanny's Voorwerp, after a Dutch schoolteacher.**

LEPUS

PASSPORT

Latin name: Lepus	**Area:** 290.3 square degrees
English name: Hare	**Number of naked-eye stars:** 73
Genitive: Leporis	**Bordering constellations:** Orion, Eridanus, Caelum, Columba, Canis Major, Monoceros
Abbreviation: Lep	
Origin: Ptolemy	**Best visibility:** November–January, south of 60° north

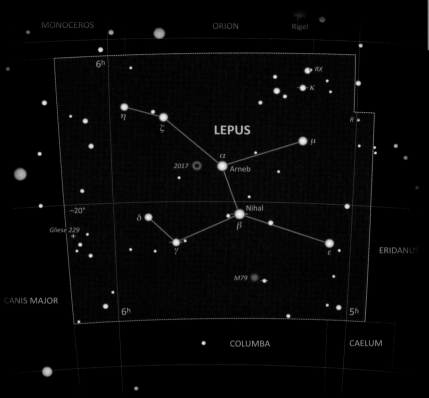

▷ **Hind's Crimson Star (R Leporis)** has an unusual red color. At maximum brightness, the star can just be seen with the naked eye.

▽ **The small white dot is the first identified brown dwarf, Gliese 229B,** which orbits the (over-exposed) red dwarf Gliese 229 (left).

LEPUS (THE HARE) is a relatively small but prominent constellation, located directly south of Orion. There are no myths associated with it, although it is sometimes described as Orion's prey.

TIMELINE

1780   French astronomer Pierre Méchain discovers the globular cluster M79, some 42,000 light-years away in Lepus. It is the best-known deep-sky object in the constellation.

1845   British astronomer John Russell Hind is the first to describe the bloodred color of the variable star R Leporis. The dying giant star, at a distance of 1,300 light-years, is also known as Hind's Crimson Star.

1995   A substellar companion is found to orbit the red dwarf star Gliese 229, which is 19 light-years away in Lepus. Gliese 229B is the first confirmed brown dwarf ever found: a "failed" star, not hot enough for spontaneous hydrogen fusion.

◀ **Old giant stars appear as orange dots in this Hubble Space Telescope image of globular cluster M79.**

LIBRA

PASSPORT

Latin name: Libra		**Area:** 538.1 square degrees	
English name: Scales		**Number of naked-eye stars:** 83	
Genitive: Librae		**Bordering constellations:** Virgo, Hydra, Centaurus (corner), Lupus, Scorpius, Ophiuchus, Serpens Caput	
Abbreviation: Lib			
Origin: Ptolemy		**Best visibility:** April–June, south of 55° north	

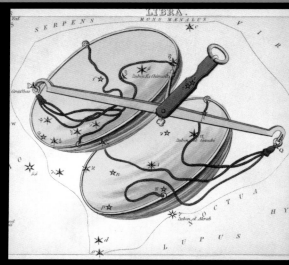

▶ In *Urania's Mirror*, an 1825 set of constellation cards, Libra is depicted with gold-covered scales.

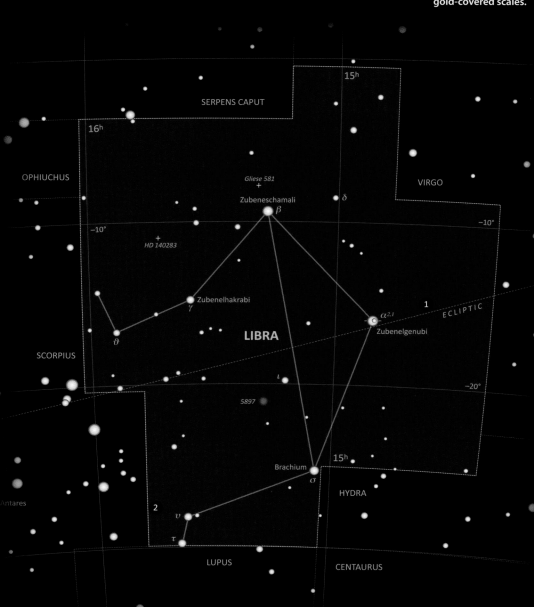
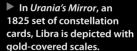

LIBRA (THE SCALES) is the only nonliving constellation of the zodiac. In ancient times, it was part of Scorpius (the Scorpion), as evidenced by the names of the two brightest stars: Zubenelgenubi (from the Arabian *Al Zuban al Janubiyyah*, or "the southern claw of the scorpion") and Zubeneschamali (the northern claw). Libra lies midway between the bright stars Spica and Antares, and is relatively easy to find, even though it lacks a really bright star or a very recognizable shape.

Libra does not contain many deep-sky objects, but it is home to exoplanet Gliese 581c—the first planet outside our solar system that was found to orbit in the so-called habitable zone of its parent star.

▶ NGC 5897 is a globular cluster in Libra, at a distance of some 24,000 light-years.

1951   A faint star in Libra, known as HD 140283, is found to have an extremely low abundance of elements heavier than hydrogen and helium, indicating a very old age. Nicknamed Methuselah, it is now regarded as the oldest known star in the Milky Way galaxy—almost as old as the universe itself.

1977   ❶ On March 10, the planet Uranus passed in front of the faint star SAO 158687 in Libra. American astronomers observed this occultation from the Kuiper Airborne Observatory and discovered the existence of a number of narrow rings around the planet. In January 1986, this ring system was imaged by the Voyager 2 spacecraft.

2005   ❷ Ground-based telescopes are aimed at Libra on July 4, where NASA's Deep Impact spacecraft releases a 820-pound "projectile" to slam into the nucleus of comet Tempel 1. The resulting outburst reveals information about the comet's internal makeup.

2007   For the first time ever, astronomers discover a planet orbiting in the habitable zone of its parent star—the region where the average temperature allows the existence of liquid water on the planet's surface. Gliese 581c is a super-Earth in a thirteen-day orbit around a faint dwarf star in Libra. Later observations reveal the planet to actually be a hothouse world like Venus. The system contains at least two more planets.

2008   A Ukrainian radio telescope is used to beam a message from Earth toward Gliese 581c. Since the planet is just over 20 light-years away, the signal won't arrive until early 2029.

◀ Stars are smeared into short trails on this long-exposure photographs of the ring system of Uranus, shot by Voyager 2 in January 1986.

▼ Just after the Deep Space Impactor slammed into the nucleus of comet Tempel 1, the mother craft captured this image of the resulting explosion.

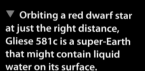

▼ Orbiting a red dwarf star at just the right distance, Gliese 581c is a super-Earth that might contain liquid water on its surface.

LUPUS

PASSPORT

Latin name: Lupus	**Area:** 333.7 square degrees
English name: Wolf	**Number of naked-eye stars:** 127
Genitive: Lupi	**Bordering constellations:** Libra, Hydra (corner), Centaurus, Circinus, Norma, Scorpius
Abbreviation: Lup	
Origin: Ptolemy	**Best visibility:** April–May, south of 30° north

LUPUS (THE WOLF) is a prominent southern constellation, lying between Scorpius (the Scorpion) and Centaurus (the Centaur). It contains a large number of bright stars, but it lacks a very characteristic shape.

TIMELINE

1006   An extremely bright supernova explodes on the border of the constellations Lupus and Centaurus, easily visible during the daytime, despite its distance of 7,200 light-years. The expanding remnant of the explosion is still visible.

2016   Using the Atacama Large Millimeter/submillimeter Array (ALMA), astronomers determine the size of dust particles surrounding the young protostar HD 142527 in Lupus. Their polarization measurements reveal that the particles are at most 150 micrometers (less than .006 inch) across.

2018   The SPHERE instrument at the European Very Large Telescope in Chile captures a detailed image of the protoplanetary disk surrounding the young variable star IM Lupi.

▼ An expanding shell of gas is all that remains from the brightest supernova in recorded history, SN 1006.

▼ Artist's impression of the dust ring surrounding the young star HD142527. The particles in the ring are smaller than 150 micrometers across.

▲ Starlight has been removed to reveal the dusty protoplanetary disk of the newborn star IM Lupi in extreme detail.

LYNX

PASSPORT

Latin name: Lynx	**Area:** 545.4 square degrees
English name: Lynx	**Number of naked-eye stars:** 97
Genitive: Lyncis	**Bordering constellations:** Camelopardalis, Auriga, Gemini, Cancer, Leo (corner), Leo Minor, Ursa Major
Abbreviation: Lyn	
Origin: Hevelius	**Best visibility:** December–February, north of 25° south

▲ A hand-colored edition of Johannes Hevelius's star atlas shows Lynx in a mirror-reversed orientation.

▶ This illustration shows how the Lynx Arc—a giant star-forming region at a distance of 12 billion light-years —may have looked when the universe was less than 2 billion years old.

LYNX IS A LARGE but obscure constellation in the northern sky. It can be found in the area between the constellations Ursa Major (the Great Bear), Auriga (the Charioteer), and Gemini (the Twins).

TIMELINE

1687 Polish astronomer Johannes Hevelius introduces the new constellation, commenting that one needs the eyesight of a lynx to actually see it.

1788 William Herschel discovers NGC 2683, sometimes called the UFO galaxy because of its shape. It lies 25 million light-years away, in the southeastern part of Lynx.

2003 Thanks to gravitational lensing by a foreground cluster, astronomers discover the Lynx Arc—a humongous star-forming region in the very early universe.

▼ NGC 2683 is seen almost edge-on, and is sometimes called the UFO galaxy.

LYRA

PASSPORT

Latin name: Lyra	**Area:** 286.5 square degrees
English name: Lyre	**Number of naked-eye stars:** 73
Genitive: Lyrae	**Bordering constellations:** Draco, Hercules, Vulpecula, Cygnus
Abbreviation: Lyr	**Best visibility:** June–July, north of 40° south
Origin: Ptolemy	

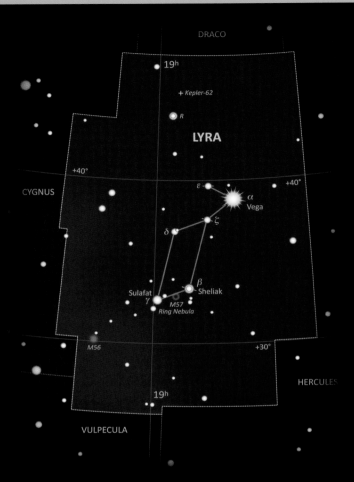

▲ M56 is a globular cluster in Lyra, discovered in 1779 by Charles Messier.

▶ Some 7,000 years ago, the star at the center of the Ring Nebula started to blow its outer layers into space.

▶ Collisions of rocky bodies may have produced the debris disk of Vega, discovered in 1984.

LYRA (THE LYRE) is one of the smaller constellations in the sky, but it is very easy to recognize as a small, slanted rectangle of stars close to brilliant Vega (Alpha Lyrae, or α Lyr), the second brightest star in the northern hemisphere. Vega is one of the vertices of the giant Summer Triangle.

Lyra is best known for the Ring Nebula (M57), a beautiful planetary nebula—the expanding shell of gas blown into space by a dying star. A small part of the constellation was also studied by NASA's Kepler space telescope; a number of exoplanets (planets orbiting other stars than our own sun) found by Kepler are located in Lyra.

▲ At 1,200 light-years from Earth, Kepler-62f is a habitable planet only slightly larger than our home world.

▼ 'Oumuamua is an elongated space rock that entered our solar system from the direction of Lyra.

TIMELINE

1779   French astronomer Antoine Darquier de Pellepoix discovers the Ring Nebula (M57), the second planetary nebula to be identified. It is about 1 light-year across and some 2,000 light-years away.

1850   American astronomers William Bond and John Whipple succeed in taking a photograph of Vega. It is the first star (other than the sun) to be photographed.

1984   The Dutch-American Infra-Red Astronomical Satellite (IRAS) observes the infrared radiation of a "debris disk" of dust and grains around Vega—one of the first such disks to be found. The discovery suggests that the star may also have a planetary system.

2013   Five planets are found to be orbiting the star Kepler-62 in Lyra. Two of them, Kepler-62e and Kepler-62f, are comparable to Earth in terms of size and composition, and orbit in the star's habitable zone, so they could potentially have oceans of liquid water.

2017   Astronomers discover the first interstellar asteroid, 'Oumuamua. The cucumber-shaped space rock appears to be coming from the direction of Lyra.

MENSΛ

PASSPORT

Latin name: Mensa	**Area:** 153.5 square degrees
English name: Table Mountain	**Number of naked-eye stars:** 22
Genitive: Mensae	**Bordering constellations:** Dorado, Hydrus, Octans, Chamaeleon, Volans
Abbreviation: Men	
Origin: de Lacaille	**Best visibility:** November–January, south of 5° north

MENSA (THE TABLE MOUNTAIN) is the faintest constellation in the sky. At 5th magnitude, even its brightest star, Alpha Mensae, or α Men, is hardly visible to the naked eye. Mensa lies very close to the south celestial pole.

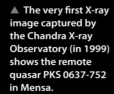

▲ **The very first X-ray image captured by the Chandra X-ray Observatory (in 1999) shows the remote quasar PKS 0637-752 in Mensa.**

◀ **NGC 1987 is a relatively compact cluster of stars within the borders of Mensa that is part of the Large Magellanic Cloud.**

▲ **The southern part of the Large Magellanic Cloud lies within the borders of Mensa. The yellow star to the lower right of the galaxy is Beta Mensae.**

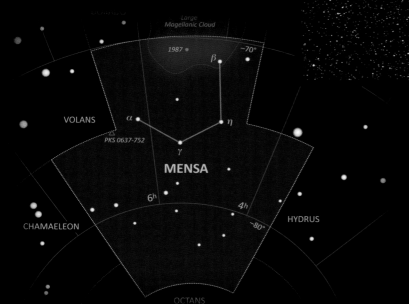

TIMELINE

1763   French astronomer Nicolas Louis de Lacaille introduces the constellation as Mons Mensae, naming it after the impressive Table Mountain, close to his observing location at Cape Town, South Africa.

1834   Also in Cape Town, John Herschel discovers a faint star cluster in Mensa, NGC 1987. It belongs to the Large Magellanic Cloud, but its true nature—globular cluster or open cluster—is not entirely clear.

1999   A quasar at 6 billion light-years from Earth, PKS 0637-752, is the first observing target for NASA's newly launched Chandra X-ray Observatory. Chandra images the X-rays of a jet, produced by a central supermassive black hole.

MICROSCOPIUM

PASSPORT

Latin name: Microscopium	**Area:** 209.5 square degrees
English name: Microscope	**Number of naked-eye stars:** 43
Genitive: Microscopii	**Bordering constellations:** Capricornus, Sagittarius, Telescopium (corner), Indus, Grus, Piscis Austrinus
Abbreviation: Mic	
Origin: de Lacaille	**Best visibility:** July–August, south of 40° north

MICROSCOPIUM (THE MICROSCOPE) is one of the fourteen obscure southern constellations introduced by Nicolas Louis de Lacaille in 1763. Lacaille named most of them after scientific instruments. It lies south of Capricornus (the Sea Goat).

▶ **This is how the dust disk of AU Microscopii might look like from a vantage point close to a hypothetical planet orbiting the red dwarf.**

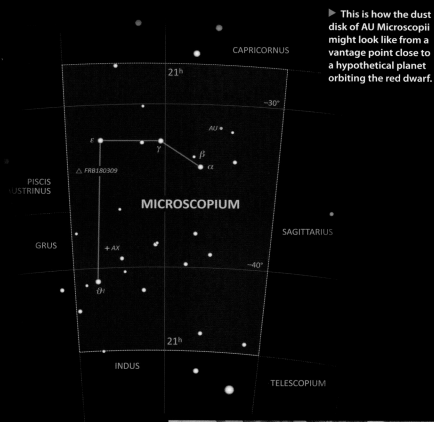

▶ **On this 1551 celestial globe, the stars that now belong to Microscopium (south of Capricornus (the Sea Goat) are not yet assigned to any constellation.**

▶ **The debris disk around the young dwarf star AU Microscopii is seen almost edge-on in this Hubble Space Telescope Image (the star itself has been blacked out).**

TIMELINE

1979 The faint red dwarf star Lacaille 8760 is found to be a flaring star, and receives a variable star designation: AX Microscopii. At just under 13 light-years away, it is the brightest red dwarf in the sky, almost but not quite visible with the naked eye.

2003 Around another red dwarf, AU Microscopii, astronomers detect a huge debris disk of dust and grains. AU Mic is just 12 million years old. So far, no planets have been found orbiting the star.

2018 Australian radio astronomers detect the most powerful fast radio burst ever, FRB 180309, within the borders of Microscopium. For about a millisecond, the mysterious source produced as much energy as a few hunderd million suns.

MONOCEROS

PASSPORT

Latin name: Monoceros

English name: Unicorn

Genitive: Monocerotis

Abbreviation: Mon

Origin: Plancius

Area: 481.6 square degrees

Number of naked-eye stars: 138

Bordering constellations: Gemini, Orion, Lepus, Canis Major, Puppis, Hydra, Canis Minor

Best visibility: December–January, between 75° north and 75° south

MONOCEROS (THE UNICORN) was introduced as a new constellation by Petrus Plancius on his 1612 celestial globe. It occupies the area between Orion and the bright stars Procyon and Sirius. However, the constellation is hard to recognize: It contains only faint stars and lacks a recognizable shape.

Since Monoceros is crossed by the Milky Way, it contains a large number of star clusters and nebulae, of which the Rosette Nebula is by far the most famous. The constellation is also home to the nearest black hole, at a "mere" 3,500 light-years from Earth.

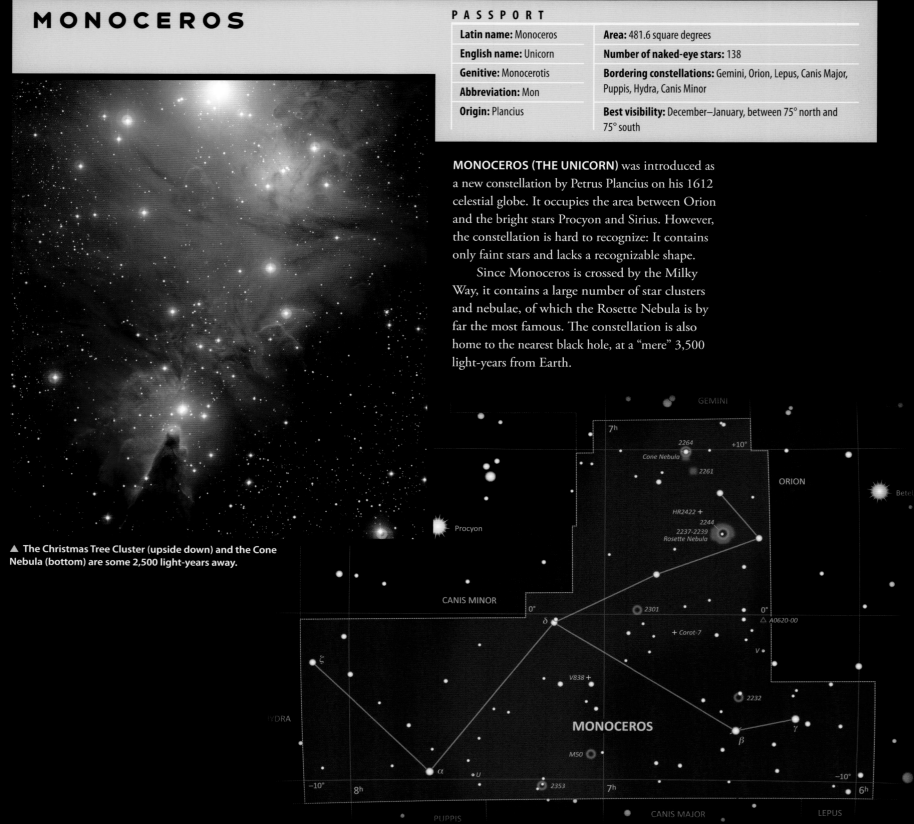

▲ The Christmas Tree Cluster (upside down) and the Cone Nebula (bottom) are some 2,500 light-years away.

◄ On the sky, the Rosette Nebula in Monoceros is more than twice the apparent diameter of the full moon.

▲ At the time of its discovery, exoplanet Corot-7b (foreground) was the smallest one known.

TIMELINE

1785 English astronomer William Herschel discovers the Cone Nebula in Monoceros—a dark dust cloud in a glowing star-forming region, close to the so-called Christmas Tree Cluster (NGC 2264).

1922 John Plaskett discovers the binary nature of the star HR 2422 in Monoceros. From its 14.4-day orbital motion, the total mass of the binary can be deduced: Plaskett's Star, as it is sometimes called, is one of the most massive binaries known, at about a hundred times the mass of the sun.

2002 In a huge explosive event, the variable star V838 Monocerotis becomes ten thousand times more luminous in just one day. Over the subsequent years, the Hubble Space Telescope images the photogenic "light echoes" of the explosion.

2009 The French satellite Corot reveals the existence of a planet just 1.6 times as large as Earth, orbiting a faint star in Monoceros now known as Corot-7. At the time of its discovery, Corot-7b was the smallest exoplanet ever found.

2018 In memory of Stephen Hawking, who died on March 14, a radio message is sent into the direction of A0620-00, the closest black hole to Earth, at 3,500 light-years in Monoceros. Orbited by a normal star, the black hole is about 10 times as massive as the sun.

▲ In 2005, three years after the giant outburst of the star V838 Mon (center), the Hubble Space Telescope captured these beautiful light echoes.

◄ A0620-00 is the closest black hole to Earth. Matter from a companion star accretes into a rotating disk before it gets devoured by the black hole.

MUSCA

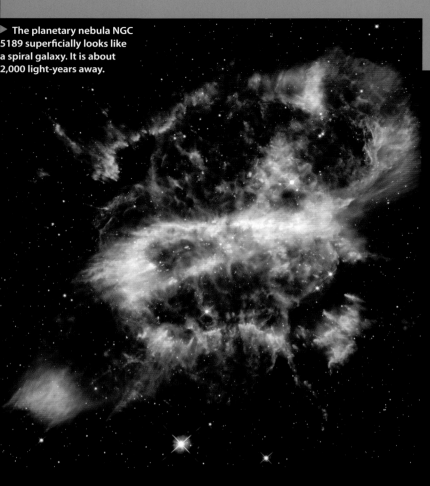

The planetary nebula NGC 5189 superficially looks like a spiral galaxy. It is about 2,000 light-years away.

PASSPORT

Latin name: Musca		**Area:** 138.4 square degrees	
English name: Fly		**Number of naked-eye stars:** 62	
Genitive: Muscae		**Bordering constellations:** Crux, Centaurus, Carina, Chamaeleon, Apus, Circinus	
Abbreviation: Mus			
Origin: Keyser and de Houtman		**Best visibility:** March–April, south of 15° north	

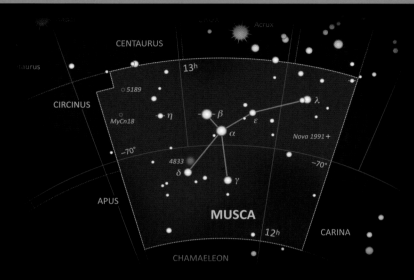

MUSCA (THE FLY) IS A SMALL CONSTELLATION close to Crux (the Southern Cross). It was named by Dutch sailors Pieter Dirkszoon Keyser and Frederick de Houtman and formally introduced by astronomer Petrus Plancius in 1598. For a long time, the constellation was also known as Apis (the Bee).

TIMELINE

1826   Scottish astronomer James Dunlop discovers NGC 5189, dubbed the Spiral Planetary Nebula. The nebula's remarkable shape is due to the presence of a binary star in its center.

1920   Around this time, Margaret Mayall and Annie Jump Cannon of the Harvard College Observatory hit upon another remarkable planetary nebula in Musca (MyCn 18), also known as the Hourglass Nebula.

1991   A nova (new star) appears in Musca. It turns out to be an explosive event in a binary system that contains a stellar-mass black hole in a 10.4-hour orbit around a normal star. Nova Muscae 1991 is also a luminous and variable source of X-rays and gamma rays.

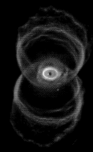

A thick equatorial dust torus may have forced the ejecta of a dying star into two giant lobes, producing the stunning Hourglass Nebula.

Nova Muscae 1991 is not yet visible in the old image at left but appears as a bright star in the image at right. The stellar explosion coincided with a luminous outburst of X-rays.

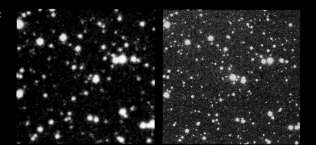

NORMA

PASSPORT

Latin name: Norma	**Area:** 165.3 square degrees
English name: Carpenter's Square	**Number of naked-eye stars:** 44
	Bordering constellations: Lupus, Circinus, Triangulum Australe, Ara, Scorpius
Genitive: Normae	
Abbreviation: Nor	**Best visibility:** April–June, south of 25° north
Origin: de Lacaille	

NORMA (THE CARPENTER'S SQUARE) is a small southern constellation, first introduced by French astronomer Nicolas Louis de Lacaille in 1763. It can be found halfway between the constellations Scorpius (the Scorpion) and Centaurus (the Centaur).

▲ Mu Normae (lower right) is one of the most luminous stars in our Milky Way galaxy. At upper left are star clusters in Scorpius.

TIMELINE

1922 American astronomer Donald Menzel discovers a highly asymmetric planetary nebula in Norma (Mz 3), nicknamed the Ant Nebula. Its shape is probably due to the fact that the nebula's central star is a binary.

1986 Astronomers discover that our Local Group of galaxies (consisting of our Milky Way galaxy, the Andromeda galaxy, the Triangulum galaxy, and dozens of dwarf galaxies) is being pulled in the direction of Norma by a giant concentration of galaxies called the Great Attractor. It is almost invisible because of dust absorption in the Milky Way, but the Norma Cluster (Abell 3627), some 220 million light-years away, is located at its core.

2007 The distance of the blue supergiant star Mu Normae, or μ Nor, is reestimated on the basis of data from the Hipparcos satellite. The new value (just over 4,000 light-years) indicates that the star is half a million times as luminous as our sun; it is one of the most luminous stars known.

▲ The Ant Nebula (Mz3) is an elongated planetary nebula at a distance of some 8,000 light-years.

▶ The Norma Cluster (Abell 3627) is at the heart of the massive Great Attractor galaxy concentration.

3D CONSTELLATIONS

CONSTELLATIONS ARE ILLUSIONARY. They don't really exist. Each and every constellation in the sky is just an *apparent* group of stars that usually are unrelated to each other. From our vantage point on Earth, we recognize the familiar shapes of the Big Dipper, the Southern Cross, and Orion. But for the hypothetical inhabitants of a planet orbiting another star in a different corner of our Milky Way galaxy, the sky looks very different. Even a mere hundred light-years from home, the view of the night sky would be completely unfamiliar to us.

The reason, of course, is that the stars that make up a particular constellation are usually at very different distances. They are distributed in three-dimensional space. As seen from Earth, the seven brightest stars in Ursa Major (the Great Bear) resemble a large celestial dipper. But seen from a different angle, they would make up a completely different pattern.

The illustration on this page shows a three-dimensional representation of the constellation Orion, the hunter. For an observer on Earth (left), the seven stars are arranged in the familiar hourglass shape, with Orion's shoulders on top, his belt in the middle, and his knees at the bottom. But as you can see, some of the stars are much farther away than others. The bright star Betelgeuse (top left) is relatively close, at some 640 light-years. The two outer stars in the belt (Alnitak and Mintaka) are much farther out: between 1,200 and 1,300 light-years. And the central star in the belt (Alnilam) is even more remote, at some 2,000 light-years. If we could look at Orion more or less from the side, as depicted in the illustration, nothing of the famous symmetrical constellation would be left.

This also means that our impression of a star's brightness is unreliable. On the sky, Betelgeuse appears brighter than Alnilam, but in reality, Alnilam's luminosity is much higher. It just appears fainter because it is farther away. Likewise, in the constellation Canis Major (the Greater Dog), Sirius appears much brighter than Wezen (Delta Canis Majoris, or δ CMa), but the truth is that Wezen produces more than three thousand times more energy. Sirius appears so bright because it is only 8.6 light-years away, while Wezen is a supergiant at a distance of some 1,600 light-years.

The same is true for every other stellar grouping. Their appearance very much depends on the fact that we observe them from the Earth. While seventeenth-century seafarers could easily navigate by observing the patterns in the night sky, future astronauts on interstellar spaceships would find no use in knowing the constellations.

Of course, the effect becomes more obvious the farther you travel away from the sun. At Proxima b, the Earth-like planet orbiting our sun's nearest neighbor Proxima Centauri, at a distance of just over 4 light-years, most constellations would still be recognizable. However, the star Alpha Centauri would appear extremely bright, and another very conspicuous star would distort the familiar W shape of the constellation Cassiopeia: our own sun.

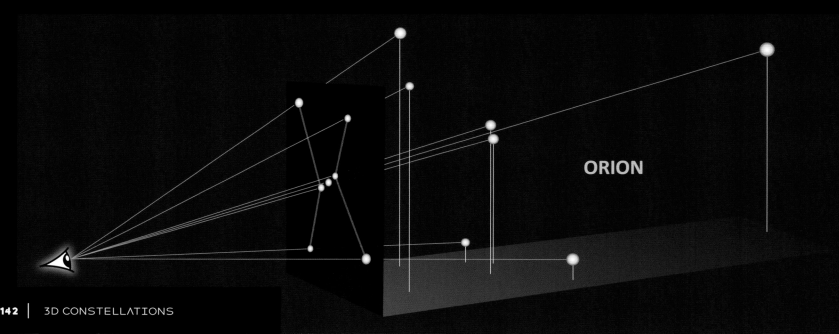

ORION

OCTANS

OCTANS (THE OCTANT) is the southernmost constellation in the sky. It contains the south celestial pole. Unlike its northern counterpart, however, the south celestial pole is not marked by a bright star. The star closest to the pole is Sigma Octantis, or σ Oct, which, at 5th magnitude, is twenty-five times fainter than the Pole Star in Ursa Minor (the Little Bear). A much easier way to locate the south celestial pole on the sky is by extending the long axis of the constellation Crux (the Southern Cross) about four times.

PASSPORT

Latin name: Octans	**Area:** 291 square degrees
English name: Octant	**Number of naked-eye stars:** 60
Genitive: Octantis	**Bordering constellations:** Pavo, Apus, Chamaeleon, Mensa, Hydrus, Tucana, Indus
Abbreviation: Oct	
Origin: de Lacaille	**Best visibility:** July–August, south of 10° south

▼ **Eighteenth-century depiction of the constellation Octans.**

TIMELINE

1730 English astronomer John Hadley invents the octant, a navigational device after which the constellation is named by Nicolas Louis de Lacaille in his 1763 southern star catalog *Coelum Australe Stelliferum*.

2005 A gas giant planet is found in an eccentric 5.2-year orbit around the brightest member of the binary star HD 142022 in Octans. The star, at 117 light-years' distance, is one of the oldest stars known that is accompanied by a planet: Its age is estimated at some 13 billion years.

▶ **The diurnal motion of the night sky above the ALMA Observatory in Chile reveals the location of the south celestial pole in Octans. However, there is no bright southern Pole Star.**

OPHIUCHUS

PASSPORT

Latin name: Ophiuchus	**Area:** 948.3 square degrees
English name: Serpent Bearer	**Number of naked-eye stars:** 174
Genitive: Ophiuchi	**Bordering constellations:** Hercules, Serpens Caput, Libra, Scorpius, Sagittarius, Serpens Cauda, Aquila
Abbreviation: Oph	
Origin: Ptolemy	**Best visibility:** May–June, between 55° north and 75° south

◄ **Thirty-five years after its launch, Voyager 1 finally left the solar system in August 2012.**

▶ **The recurrent nova RS Ophiuchi consists of a red giant and a white dwarf, orbiting each other every 454 days.**

OPHIUCHUS (THE SERPENT BEARER) is one of the larger constellations in the sky, located between Scorpius (the Scorpion) and Aquila (the Eagle). It represents the half-god Asclepius, son of Apollo. After the death of his mother, Koronis, Asclepius was raised by the noble centaur Chiron, who introduced him into the art of healing. However, Zeus became afraid that humans would become immortal because of the healing powers of Asclepius, so he killed him with a lightning bolt.

The brightest star in the constellation, Rasalhague (Alpha Ophiuchi, or α Oph) represents the healer's head. In the sky, it is very close to Rasalgethi, the head of Hercules. Ophiuchus contains a large number of globular clusters, star-forming regions and dark nebulae. The constellation also served as the backdrop for quite a number of highlights in planetary exploration.

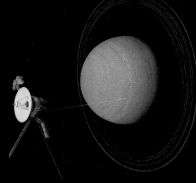

▲ **Artist's impression of Voyager 2's encounter with the highly tilted planet Uranus.**

▼ **This multi-wavelength view of Kepler's supernova remnant combines optical, infrared, and X-ray images of the expanding shell of hot gas.**

▶ **M10 is one of the nearest globular clusters in our Milky Way galaxy.**

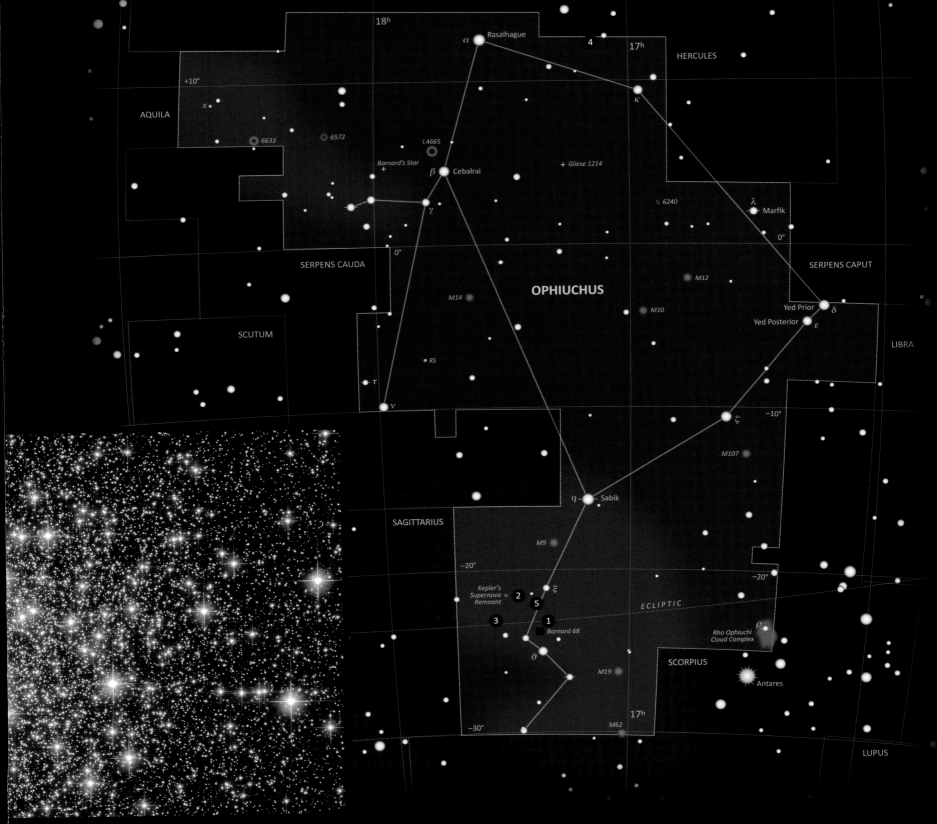

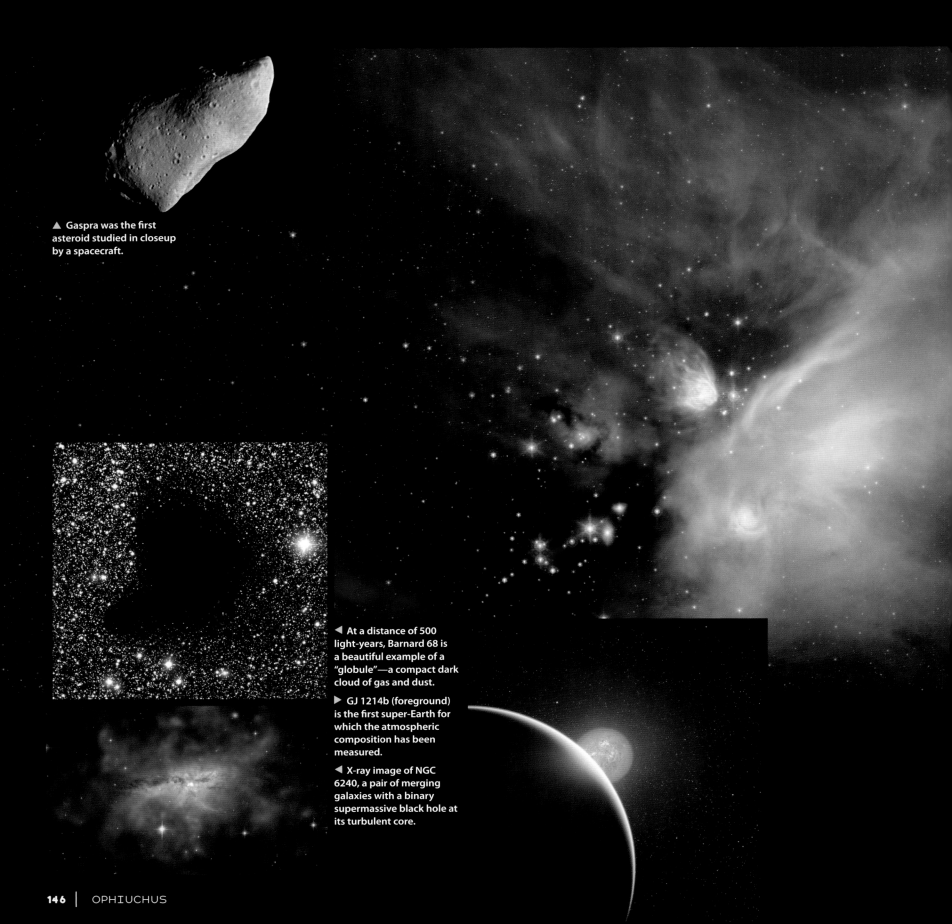

▲ Gaspra was the first asteroid studied in closeup by a spacecraft.

◀ At a distance of 500 light-years, Barnard 68 is a beautiful example of a "globule"—a compact dark cloud of gas and dust.

▶ GJ 1214b (foreground) is the first super-Earth for which the atmospheric composition has been measured.

◀ X-ray image of NGC 6240, a pair of merging galaxies with a binary supermassive black hole at its turbulent core.

TIMELINE

1604 Johannes Kepler discovers a bright supernova in Ophiuchus. For a few weeks, it is even visible in bright daylight. The stellar explosion, the remnant of which can still be seen, occurred at a distance of some 20,000 light-years.

1764 French astronomer Charles Messier comes across a number of globular clusters in the constellation: M9, M10, M12, M14, and M19. At some 15,000 light-years, M10 is relatively close.

1898 For the first time, an outburst is seen from the star RS Ophiuchi. This "recurrent nova" consists of a red giant and a white dwarf orbiting each other. Gas from the red giant star accretes on the surface of the white dwarf, producing intermittent thermonuclear explosions.

1916 Edward Emerson Barnard discovers the fastest-moving star in the sky. Barnard's Star, just 6 light-years away in Ophiuchus (it is the nearest star except for the three stars in the Alpha Centauri system), covers a distance equivalent to the apparent diameter of the moon in just 175 years.

1919 Barnard also discovers a compact dark cloud, now known as Barnard 68, in Ophiuchus. Half a light-year across and twice the mass of the sun, it may collapse into a star in the future.

1986 ❶ NASA's Voyager 2 spacecraft encounters Uranus, on January 24. At the time, the giant planet is crossing the southern part of Ophiuchus. Uranus turns out to be a very bland planet, surrounded by dark, narrow rings and accompanied by a retinue of heavily battered moons.

◀ The Cassini spacecraft flies between Saturn's rings and cloud tops, just before it plunges into the planet's atmosphere.

1991 ❷ The asteroid Gaspra is in Ophiuchus when the Galileo spacecraft makes a flyby at 1,600 kilometers distance on October 29—the first asteroid encounter ever.

1995 ❸ On December 7, Galileo arrives at the giant planet Jupiter, while Jupiter is within Ophiuchus. The spacecraft's atmospheric probe descends into Jupiter's atmosphere and carries out the first in situ measurements of its composition.

2002 Studies by NASA's Chandra X-ray Observatory reveal that the merging galaxy pair NGC 6240, at a distance of 400 million light-years, harbors a binary supermassive black hole. At present, the two black holes are some 3,000 light-years apart, but within a billion years or so, they will collide and merge.

2009 A small, low-density planet is found orbiting the red dwarf star Gliese 1214, just 40 light-years away in Ophiuchus. The planet, known as GJ 1214b, may be surrounded by a very thick atmosphere of water vapor. It may even be a "water world," covered by a planet-wide ocean.

2012 ❹ At the northern edge of Ophiuchus, Voyager 1, launched in September 1977, is leaving the solar system when it crosses the so-called heliopause—the edge of the sun's magnetic sphere of influence.

2012 For the first time, sugar molecules (glycolaldehyde) are discovered in interstellar space, in the active and nearby Rho Ophiuchi star-forming cloud complex.

2017 ❺ Saturn is located in the southern part of Ophiuchus when, on September 15, the Cassini spacecraft makes its kamikaze dive into the planet's atmosphere.

ORION

PASSPORT

Latin name: Orion	**Area:** 594.1 square degrees
English name: Orion	**Number of naked-eye stars:** 204
Genitive: Orionis	**Bordering constellations:** Taurus, Eridanus, Lepus, Monoceros, Gemini
Abbreviation: Ori	
Origin: Ptolemy	**Best visibility:** November–January, between 75° north and 65° south

ORION IS ONE OF THE MOST recognizable constellations in the night sky, thanks to its bright stars and its conspicuous hourglass shape. The three central stars—Orion's Belt—are located almost exactly on the celestial equator and can be seen from all over the world, except for the north and south poles. Orion is situated midway between the bright stars Aldebaran and Sirius.

In Greek mythology, Orion was the handsome son of the sea god Poseidon and princess Euryale, daughter of King Minos of Crete. He became an accomplished hunter; Canis Major (the Greater Dog) and Canis Minor (the Lesser Dog) are Orion's hunting dogs. He is most often depicted as raising a shield and a club to attack Taurus (the Bull).

By far the most famous deep sky object in Orion is the Orion Nebula, a large star-forming region just south of the three Belt stars. It can be glimpsed with the naked eye as a bit of nebulosity surrounding the small star cluster Theta Orionis, or ϑ Ori.

▶ In his 1627 star atlas *Coelum Stellatum Christianum*, Julius Schiller depicted Orion as Saint Joseph, carpenter and father of Jesus.

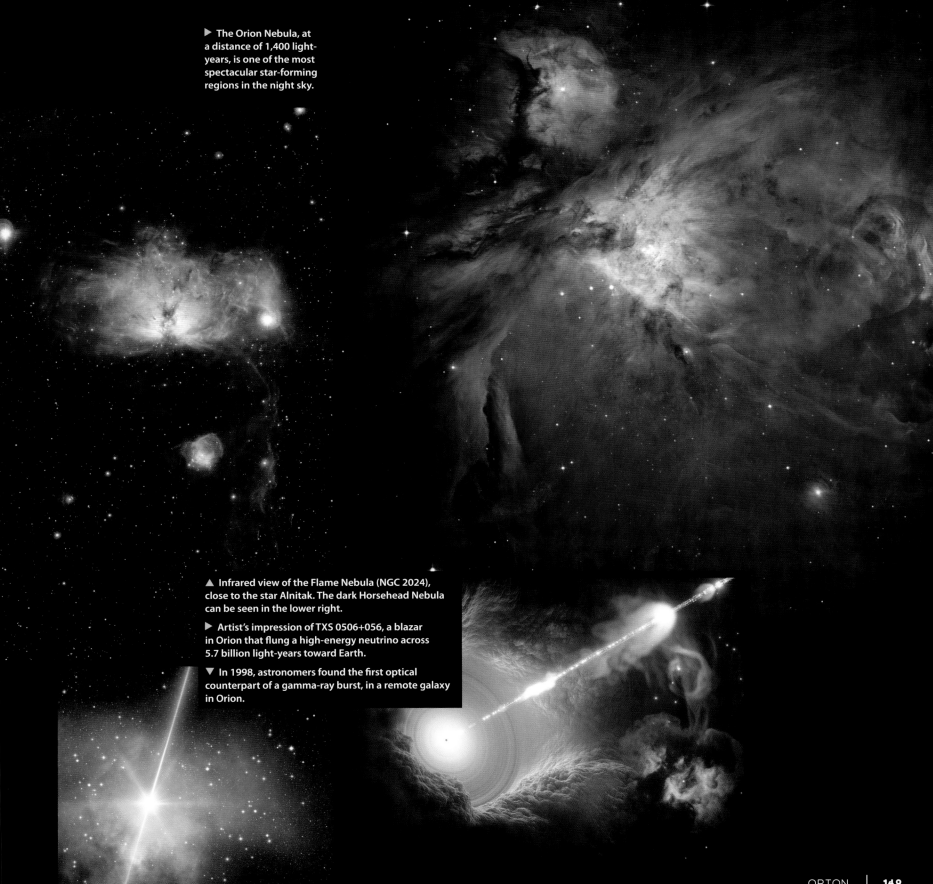

▶ The Orion Nebula, at a distance of 1,400 light-years, is one of the most spectacular star-forming regions in the night sky.

▲ Infrared view of the Flame Nebula (NGC 2024), close to the star Alnitak. The dark Horsehead Nebula can be seen in the lower right.

▶ Artist's impression of TXS 0506+056, a blazar in Orion that flung a high-energy neutrino across 5.7 billion light-years toward Earth.

▼ In 1998, astronomers found the first optical counterpart of a gamma-ray burst, in a remote galaxy in Orion.

1610 French astronomer Nicolas-Claude Fabri de Peiresc is the first one to describe the faint nebulosity in Orion's Sword—the Orion Nebula, also known as M42. Apart from the Tarantula Nebula in the Large Magellanic Cloud, this is the brightest star-forming region in the sky. It lies at a distance of some 1,400 light-years and measures 25 light-years across.

1617 Galileo Galilei discovers, in the brightest part of the Orion Nebula, a small group of bright and young stars that are close together on the sky and in space. The brightest star in the small cluster (now known as the Trapezium) has an extremely high surface temperature of 45,000 degrees F.

1831 German astronomer Friedrich von Struve discovers that the bright star Rigel (Beta Orionis, or β Ori) is actually a binary star. While the main star is a blue supergiant, eighty-five thousand times more luminous than the sun, its companion is much fainter and was later found to be a binary itself. Rigel is some 770 light-years from Earth.

1888 At Harvard College Observatory, Williamina Fleming discovers the Horsehead Nebula (also known as Barnard 33)—a protruding dark cloud of dust, silhouetted against the bright glow of the emission nebula IC 434, just south of the star Alnitak (Zeta Orionis, or ζ Ori).

1894 On long-exposure photographs of Orion, Edward Emerson Barnard discovers a huge, almost circular loop of faint nebulosity east of Orion's Belt. Barnard's Loop is some 300 light-years across and was probably formed by a supernova explosion some 2 million years ago.

1920 Through a technique known as interferometry, Albert Michelson and Francis Pease succeed in measuring the angular diameter of Betelgeuse (Alpha Orionis, or α Ori)—an astronomical first. The orange subgiant, which will probably explode as a supernova within a million years or so, turns out to be almost 400 million kilometers in diameter.

1993 The Hubble Space Telescope discovers protoplanetary disks (proplyds) around a number of young stars in the Orion Nebula. It is the first time that such circumstellar disks of gas and dust are actually imaged around newborn stars.

1997 On February 28, Dutch astronomers Titus Galama and Paul Groot identify the optical counterpart of a gamma-ray burst (GRB 970228) that was detected that same day by the Italian-Dutch satellite BeppoSAX. The find, in the northwestern part of Orion, proves that gamma-ray bursts are extremely energetic events in remote galaxies.

2017 For the first time ever, the origin of a high-energy cosmic neutrino is determined. The source, TXS 0506+056, at some 4 billion light-years away, is a blazar—a galaxy with a central black hole that fires a jet of energetic particles in our direction.

◀ **M78 is a large emission nebula in Orion, dissected by dark bands of absorbing dust.**

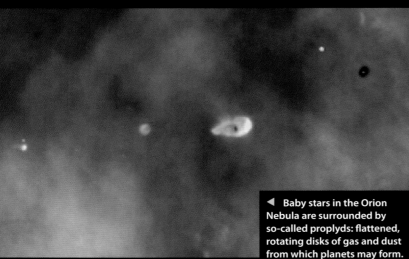

◀ **Baby stars in the Orion Nebula are surrounded by so-called proplyds: flattened, rotating disks of gas and dust from which planets may form.**

◄ Barnard's Loop runs around the Belt and Sword of Orion. Betelgeuse is at upper left, Rigel at lower right.

▲ The European Very Large Telescope took this color image of the aptly-named Horsehead Nebula in Orion.

► The Hubble Space Telescope's infrared NICMOS camera captured this image of the four bright, young stars in the Trapezium star cluster.

PAVO

PASSPORT

Latin name: Pavo	**Area:** 377.7 square degrees
English name: Peacock	**Number of naked-eye stars:** 87
Genitive: Pavonis	**Bordering constellations:** Telescopium, Ara, Apus, Octans, Indus
Abbreviation: Pav	**Best visibility:** June–August, south of 15° north
Origin: Keyser and de Houtman	

PAVO (THE PEACOCK) is a constellation in the southern sky that was first introduced by Dutch seafarers Pieter Dirkszoon Keyser and Frederick de Houtman. Keyser and de Houtman named quite a few constellations after exotic creatures they encountered during their travels. Based on their descriptions, astronomer Petrus Plancius incorporated Pavo on a celestial globe in 1598.

The constellation sits in a relatively empty part of the sky, but the brightest star, officially named Peacock (Alpha Pavonis, or α Pav), is very prominent. It marks the head of the bird, and can be found between the constellation Capricornus (the Sea Goat) and the south celestial pole.

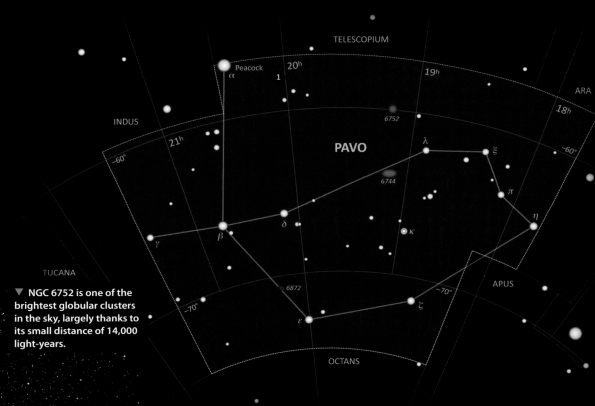

▼ **NGC 6752 is one of the brightest globular clusters in the sky, largely thanks to its small distance of 14,000 light-years.**

◄ **Pavo and Indus appear in the 1742 *Atlas Coelestis* by Johann Gabriel Doppelmayr.**

TIMELINE

1826   Scottish astronomer James Dunlop, working in South Africa, discovers the globular cluster NGC 6752 in Pavo. Measuring about 100 light-years across, this is the third-brightest globular cluster in the sky.

1835   From his observatory near Cape Town, South Africa, English astronomer John Herschel (son of William Herschel, the discoverer of Uranus) studies the southern sky and discovers many nebulae and star clusters, including NGC 6872 in Pavo. This galaxy, nicknamed the Condor galaxy, is one of the largest barred spiral galaxies known, measuring some 380,000 light-years across. Its shape is distorted by tidal forces from a nearby elliptical galaxy, IC 4970.

1977   ❶ On August 20, NASA's Voyager 2 spacecraft is launched from Cape Canaveral, Florida. It is the first spacecraft to study four planets in succession: Jupiter (in 1979), Saturn (1981), Uranus (1986), and Neptune (1989). It is now heading out of our solar system, in the direction of the constellation Pavo.

2003   The star Delta Pavonis, or δ Pav, at 20 light-years from Earth, is listed as the most promising SETI target (Search for Extraterrestrial Intelligence) among the one hundred nearest sun-like stars. So far, however, no planets have been discovered orbiting the star.

◀ The star Delta Pavonis (upper left) is very similar to our own sun.
▼ The giant Condor galaxy (NGC 6872) is disturbed by the gravity of its companion, IC 4970 (top).

▶ Voyager 2 is leaving the solar system in the direction of the constellation Pavo.

PEGASUS

PASSPORT

Latin name: Pegasus	**Area:** 1,120.8 square degrees
English name: Pegasus	**Number of naked-eye stars:** 177
Genitive: Pegasi	**Bordering constellations:** Lacerta, Cygnus, Vulpecula, Delphinus, Equuleus, Aquarius, Pisces, Andromeda
Abbreviation: Peg	
Origin: Ptolemy	**Best visibility:** August–October, north of 50° south

▼ The Propeller galaxy (NGC 7479) has a radio-emitting jet that spins in the wrong direction, maybe due to a past merger with a smaller galaxy.

PEGASUS IS A VERY LARGE AND PROMINENT constellation in the northern autumn sky, representing a winged horse. It is easy to recognize: The horse's body is marked by four bright stars, arranged in a large square. (Actually, the star in the northeastern corner of this Great Square of Pegasus officially belongs to the neighboring constellation Andromeda.)

In Greek mythology, Pegasus was born out of the blood of the monstrous Medusa, who got beheaded by Perseus just before he went on to rescue Andromeda. Pegasus was also the horse of Bellerophon, the hero who killed the heinous chimaera before he (unsuccessfully) tried to fly to Olympus, the mountain of the gods.

It was within the borders of Pegasus that the first planet orbiting a sun-like star was found. Apart from a couple of exoplanet firsts, Pegasus also contains a number of interesting galaxies and other deep-sky objects.

▼ Hubble Space Telescope image of the globular cluster M15 in Pegasus.

b

c

d

▲ Except for NGC 7320 (top center), the galaxies in Stephan's Quintet are really close together in space and will collide and merge in the future.

◄ Three planets are seen to orbit the star HR 8799. The colored blob in the center is the result of artificially obscuring the light of the star.

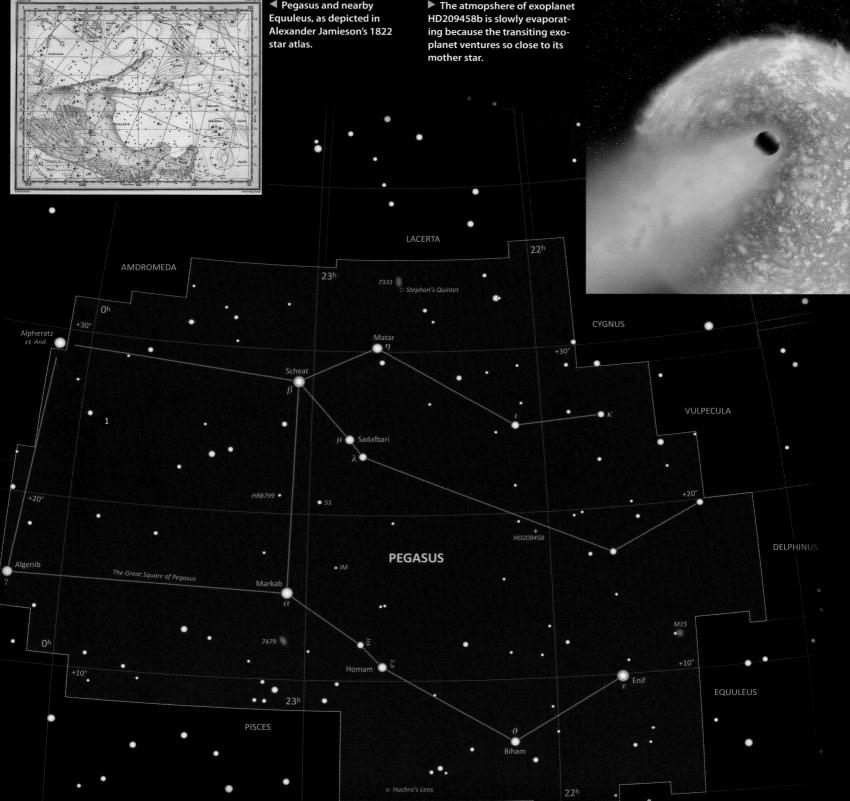

◄ **Pegasus and nearby Equuleus, as depicted in Alexander Jamieson's 1822 star atlas.**

▶ **The atmopshere of exoplanet HD209458b is slowly evaporating because the transiting exoplanet ventures so close to its mother star.**

LACERTA

22ʰ

23ʰ

7331 ○ Stephan's Quintet

AMDROMEDA

0ʰ

+30°

CYGNUS

Alpheratz
α And

+30°

Matar
η

Scheat
β

ι

κ

VULPECULA

μ Sadalbari
λ

1

+20°

HR8799

51

+20°

+
HD209458

DELPHINUS

PEGASUS

Algenib
γ

The Great Square of Pegasus

IM

M15

Markab
α

7479

ξ

+10°

0ʰ

Homam ζ

Enif
ε

+10°

23ʰ

EQUULEUS

PISCES

ϑ

Biham

⊛ Huchra's Lens

22ʰ

1746 In France, Jean-Dominique Maraldi discovers the beautiful globular cluster M15, not too far from Enif (Epsilon Pegasi or ε Peg), the brightest star in the constellation. M15 is 34,000 light-years away and was the first globular cluster found to contain a planetary nebula (in 1928).

1784 In England, William Herschel discovers a number of galaxies in Pegasus, including the Propeller galaxy (NGC 7479) and the beautiful NGC 7331, one of the first galaxies that was found to display a spiral pattern.

1877 French astronomer Édouard Stephan finds a very compact group of five interacting galaxies in Pegasus, now known as Stephan's Quintet. One of them, NGC 7320, later turned out to be a foreground object at 40 million light-years, but the other four, seven times farther away, are really close together and will eventually merge into one large elliptical galaxy.

1985 American astronomer John Huchra discovers a gravitational lens in Pegasus. Huchra's Lens consists of four images of the same remote quasar (8 billion light-years away), created by the gravitational light-bending effect of a foreground galaxy at 400 million light-years. The configuration is also known as the Einstein Cross.

1995 On October 6, Swiss astronomers Michel Mayor and Didier Queloz announce the discovery of 51 Pegasi b, the first planet found around a sun-like star. The Jupiter-like world completes one orbit every four days. The star, at 51 light-years from Earth, is just visible with the naked eye.

1999 Another exoplanet in Pegasus, HD 209458b, turns out to transit across the face of its parent star every 3.5-day orbit, creating minute periodic dips in the star's brightness. This marks the first time an exoplanet is detected through the transit method. Later observations reveal that the planet's outer gas layers are slowly evaporating under the fierce radiation of the nearby star.

2004 NASA launches its Gravity Probe B spacecraft, designed to test Einstein's general theory of relativity from space. The satellite's telescope is continuously aimed at the star IM Pegasi, the position of which is known with great accuracy from radio observations.

2008 For the first time ever, the planetary system of another star is directly imaged (at infrared wavelengths), by the Keck and Gemini telescopes at Mauna Kea, Hawaii. The 30-million-year-old star HR 8799 in Pegasus, at a distance of 130 light-years, is accompanied by at least four large planets in wide, circular orbits.

2017 ❶ Astronomers discover the first interstellar visitor to the solar system: an elongated object called 'Oumuamua. After a brief—and fast—flyby of the sun, it disappears from view and is now heading into deep space in the direction of the constellation Pegasus.

▶ IM Pegasi (upper right) was the target of NASA's Gravity Probe B spacecraft. The bright star at lower left is Markab (Alpha Pegasi).

▼ The gravity field of a foreground galaxy (center) splits the light from a background quasar into four separate images, creating the famous Einstein Cross.

▼ 51 Pegasi b (left) was the first exoplanet to be discovered orbiting a sun-like star.

▼ NGC 7331, some 45 million light-years away in Pegasus, is very similar to our own Milky Way galaxy.

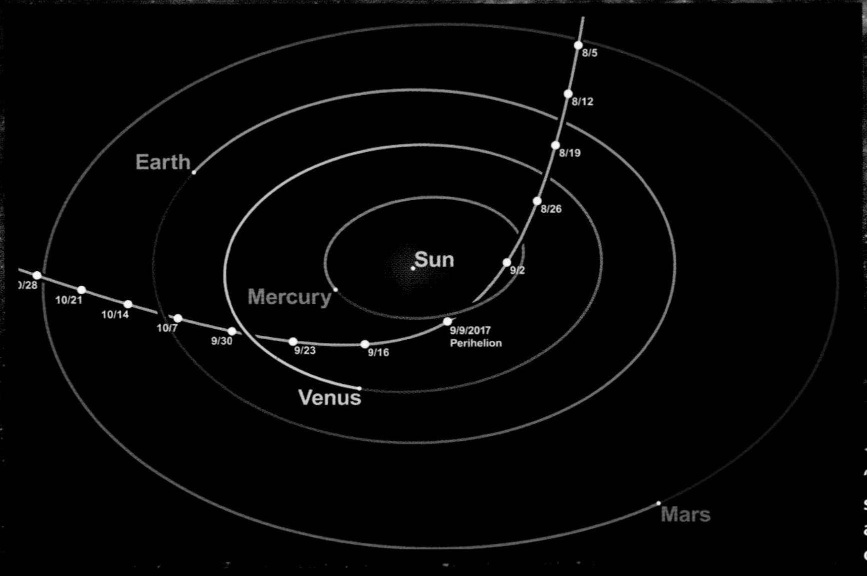

Earth

Sun

Mercury

Venus

Mars

8/5

8/12

8/19

8/26

9/2

9/9/2017 Perihelion

9/16

9/23

9/30

10/7

10/14

10/21

)/28

◀ The hyperbolic path of 'Oumuamua through the inner solar system, before it sped away in the direction of the constellation Pegasus.

PERSEUS

PASSPORT

Latin name:	Perseus	**Area:**	615.0 square degrees
English name:	Perseus	**Number of naked-eye stars:**	158
Genitive:	Persei	**Bordering constellations:** Cassiopeia, Andromeda, Triangulum, Aries, Taurus, Auriga, Camelopardalis	
Abbreviation:	Per		
Origin:	Ptolemy	**Best visibility:** October–December, north of 30° south	

◀ **The famous Double Cluster in Perseus, also known as h and χ Persei, lies at a distance of over 7,000 light-years, but is easily visible to the naked eye.**

▼ **Dusty filaments are silhouetted against the bright glow of the active galaxy NGC 1275, at the heart of the Perseus cluster.**

▶ **Multi-wavelength view of GK Persei (Nova Persei 1901). Extremely hot debris from the explosion is still expanding at 300 kilometers per second.**

PERSEUS IS A WELL-KNOWN CONSTELLATION in the northern sky, lying between Cassiopeia and Taurus (the Bull). It is crossed by the central plane of our Milky Way galaxy. As a result, it contains a large number of star clusters and nebulae.

Perseus is one of the protagonists of the myth of Andromeda, daughter of Cassiopeia and Cepheus. After Andromeda was chained to a rock cliff as a sacrifice to the sea monster Cetus, she was rescued by Perseus, who had first slain Medusa.

The star that marks the head of Medusa, Algol (Beta Persei, or β Per) is the best-known variable star in the sky. Its inconstant nature was already known to Egyptian sky watchers, long before the invention of the telescope.

▼ Infrared image of the California Nebula in Perseus, discovered by Edward Emerson Barnard in 1884.

◀ An artist's impression of the superluminous hypernova SN 2006gy, which exploded in galaxy NGC 1260.

TIMELINE

1782   John Goodricke is the first to note the regularity in Algol's brightness variations, which had been described by Geminiano Montanari in 1667. Goodricke also has an explanation for the star's behavior: It is orbited by a cooler, darker companion that periodically blocks some of its light.

1862   Lewis Swift and Horace Tuttle discover the comet that is named after them. Dust particles in the comet's 133-year orbital path enter Earth's atmosphere every year around August 12, as part of the well-known Perseid meteor shower, named after the constellation from which the "shooting stars" appear to originate.

1901   On February 21, Scottish amateur astronomer Thomas Anderson discovers a bright nova in Perseus. For a short while, Nova Persei 1901 was one of the brightest stars in the sky. Also known as GK Persei, the star still displays regular smaller outbursts.

2006   In NGC 1260, a galaxy at a distance of 240 million light-years, astronomers witness one of the first hypernovas ever observed. SN 2006gy was about a hundred times more energetic than a regular supernova.

2013   Detailed measurements by the Gemini North Telescope on Mauna Kea, Hawaii, reveal that the supermassive black hole in the core of NGC 1275—the central galaxy of the massive Perseus Cluster—is 800 million times as massive as the sun. However, later estimates arrive at a value of just 30 million solar masses.

PHOENIX

PASSPORT

Latin name: Phoenix	**Area:** 469.3 square degrees
English name: Phoenix	**Number of naked-eye stars:** 71
Genitive: Phoenicis	**Bordering constellations:** Sculptor, Grus, Tucana, Hydrus (corner), Eridanus, Fornax
Abbreviation: Phe	
Origin: Keyser and de Houtman	**Best visibility:** September–October, south of 30° north

◄ The small circle marks the X-ray source HLX-1—an intermediate-mass black hole in the remote galaxy ESO 243-49.

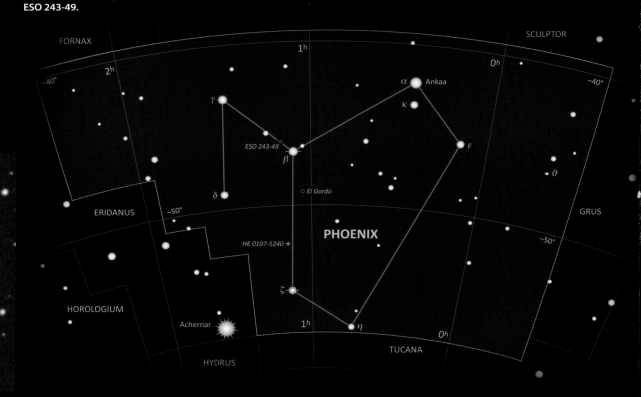

▲ The arrow points to one of the oldest stars in our Milky Way galaxy, with an age of 13 billion years or so.

▼ Blue colors denote the distribution of dark matter in the massive galaxy cluster El Gordo, which is 3,000 times the mass of our Milky Way galaxy.

PHOENIX IS ONE OF the southern constellations conceived by Dutch sailors Pieter Dirkszoon Keyser and Frederick de Houtman, and formally introduced by astronomer Petrus Plancius in 1598. It can be found close to the bright star Achernar.

TIMELINE

2002 Astronomers discover a star at 36,000 light-years' distance in Phoenix that contains extremely low quantities of elements heavier than hydrogen and helium. This suggests that the star (HE 0107-5240) is almost as old as the universe itself.

2009 An intermediate-mass black hole is found in the galaxy ESO 243-49 in Phoenix, at 290 million light-years from Earth. The X-ray source (HLX-1) turns out to be at the core of a small star cluster.

2012 Astronomers find one of the most massive galaxy clusters in the early universe at a distance of 7.2 billion light-years. Officially known as ACT-CL J0102-4915, the giant cluster is nicknamed El Gordo (the fat one).

PICTOR

PASSPORT

Latin name: Pictor	**Area:** 246.7 square degrees
English name: Painter	**Number of naked-eye stars:** 49
Genitive: Pictoris	**Bordering constellations:** Columba, Caelum, Dorado, Volans, Carina, Puppis
Abbreviation: Pic	
Origin: de Lacaille	**Best visibility:** November–January, south of 25° north

▲ **Kapteyn b (left) is larger and more massive than the Earth. It is the oldest-known potentially habitable exoplanet.**

PICTOR (THE PAINTER) is a rather obscure galaxy on the southern sky, between the bright star Canopus and the Large Magellanic Cloud. It was introduced by Nicolas Louis de Lacaille in 1763.

▼ **Flattened by its fast rotation, exoplanet Beta Pictoris b orbits well within the star's flattened debris disk.**

TIMELINE

1898 Dutch astronomer Jacobus Kapteyn discovers a red dwarf at 12.8 light-years away in Pictor with an extremely fast apparent motion across the sky. Kapteyn's Star probably escaped from the globular cluster Omega Centauri.

1983 The Dutch-American Infra-Red Astronomical Satellite (IRAS) detects the infrared radiation from a dusty disk around the star Beta Pictoris, or β Pic—the first debris disk ever discovered. The star is also accompanied by at least one giant planet, discovered in 2008.

2014 Two super-Earths are found to orbit Kapteyn's Star. One of them is in the habitable zone of the red dwarf star, where temperatures allow the existence of oceans. At 11 billion years—the age of the system—this is the oldest-known habitable exoplanet.

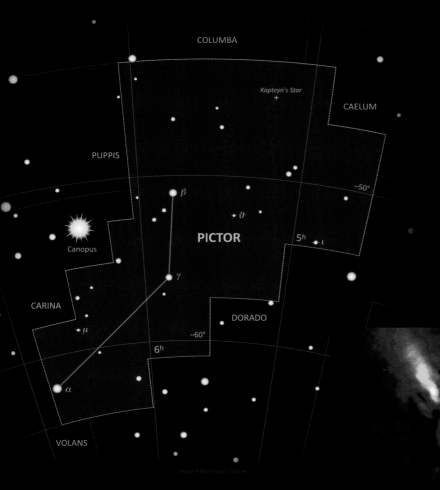

◄ **By blocking the light of the star, the infrared radiation of Beta Pictoris's dusty debris disk, seen edge-on from Earth, becomes visible.**

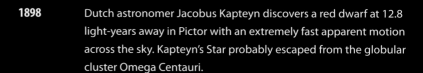

PISCES

PASSPORT

Latin name: Pisces	**Area:** 889.4 square degrees
English name: Fishes	**Number of naked-eye stars:** 150
Genitive: Piscium	**Bordering constellations:** Andromeda, Pegasus, Aquarius, Cetus, Aries, Triangulum
Abbreviation: Psc	
Origin: Ptolemy	**Best visibility:** September–October, between 80° north and 55° south

PISCES (THE FISHES) is one of the largest constellations of the zodiac, and one of the oldest in the sky, probably dating back to Sumerian times. It can be found southeast of the Great Square of Pegasus. Its lack of bright stars makes it hard to recognize from urban areas.

The two fishes represent Aphrodite and Eros, who dived into the sea and transformed themselves into fishes to escape the monster Typhon.

To astronomers, the constellation is known because it contains the vernal equinox—the point in the sky where the sun crosses the celestial equator from south to north every year around March 20.

▲ Pink regions in the spiral arms of M74 are clouds of glowing hydrogen in which new stars are born.

TIMELINE

1780   French astronomer Pierre Méchain discovers the face-on spiral galaxy M74, at a distance of some 32 million light-years.

1917   Adriaan van Maanen, a Dutch astronomer, discovers a white dwarf star in Pisces, at just 14 light-years from Earth. Van Maanen's star is the nearest known white dwarf that is not part of a binary system.

1976   A faint dwarf galaxy is found in Pisces by Russian astronomer Valentina Karachentseva. The Pisces dwarf is a satellite galaxy of the Triangulum galaxy (M33), some 2.5 million light-years away.

2002   A hypernova—an unusually bright supernova—is discovered in M74 by Yoji Hirose, a Japanese amateur astronomer.

▼ The Pisces dwarf galaxy is a very faint companion to M33, the large spiral galaxy in nearby Triangulum (the Triangle).

◄ The galaxy cluster CL0024+1654 in Pisces, some 3.6 billion light-years away, contains large amounts of dark matter. Its distibution, derived from the gravitational lens effect on background galaxies, is represented in blue.

▶ Apart from binary companions Sirius B and Procyon B, Van Maanen's Star (center) is the nearest white dwarf in our Milky Way galaxy.

TRIANGULUM

1ʰ

+30°

τ

υ

φ

ANDROMEDA

Pisces Dwarf

ψ¹

+20°

+20°

ARIES

M74

η

CL 0024+1654

PISCES

PEGASUS

+10°

2ʰ

o

+10°

23ʰ

ζ

ε

δ

ω

ϑ

ι

The Circlet

β

ν

TX

γ

α

Van Maanen's Star

λ

κ

0°

Arescha

1ʰ

ECLIPTIC

CETUS

March
Equinox

0°

0°

23ʰ

AQUARIUS

0ʰ

PISCIS AUSTRINUS

PASSPORT

Latin name: Piscis Austrinus	**Area:** 245.4 square degrees
English name: Southern Fish	**Number of naked-eye stars:** 47
Genitive: Piscis Austrini	**Bordering constellations:** Aquarius, Capricornus, Microscopium, Grus, Sculptor
Abbreviation: PsA	
Origin: Ptolemy	**Best visibility:** August–September, south of 50° north

PISCIS AUSTRINUS (THE SOUTHERN FISH) is a small constellation south of Aquarius (the Water Bearer). On ancient decorative charts, the fish is drinking the water poured by Aquarius. Piscis Austrinus is mainly known for its brightest star, Fomalhaut (Alpha Piscis Austrini, or α PsA), which is the eighteenth brightest star in the night sky.

The name Fomalhaut derives from the Arabic *fum al-ḥawt* (mouth of the fish). The star is about twice as large and twice as massive as our own sun, but it has almost seventeen times the sun's luminosity. It is also relatively close—some 25 light-years away.

Fomalhaut's age is less than half a million years. The star is surrounded by a flattened, rotating disk of gas and dust in which at least one giant planet is forming.

Apart from Fomalhaut, Piscis Austrinus contains few interesting objects, which is of course mainly due to the constellation's small size.

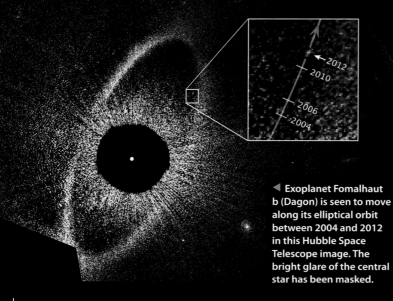

◀ **Exoplanet Fomalhaut b (Dagon) is seen to move along its elliptical orbit between 2004 and 2012 in this Hubble Space Telescope image. The bright glare of the central star has been masked.**

▲ **Fomalhaut is the brightest star in Piscis Austrinus, and the eighteenth brightest star in the night sky.**

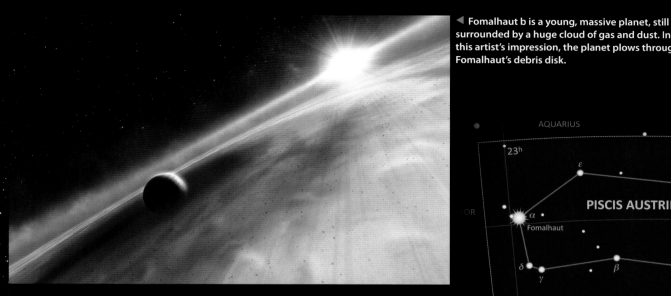

◀ Fomalhaut b is a young, massive planet, still surrounded by a huge cloud of gas and dust. In this artist's impression, the planet plows through Fomalhaut's debris disk.

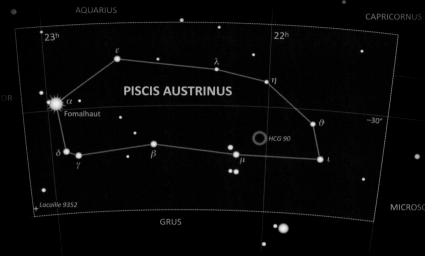

TIMELINE

1834   In South Africa, John Herschel discovers three galaxies in Pisces that are very close together in the sky: NGC 7173, NGC 7174, and NGC 7176. Astronomers have since realized that they are also close together in space, and that they interact with each other gravitationally—the trio is known as Hickson Compact Group 90 (HCG 90).

1881   Benjamin Gould notes the high proper motion (apparent motion across the sky) of the faint red dwarf star Lacaille 9352. The star itself had earlier been found by French astronomer Nicolas Louis de Lacaille in the extreme southeast corner of Piscis Austrinus. At 10.75 light-years, it is the twelfth-closest star system to Earth. Its proper motion amounts to 6.9 arc seconds per year—the fourth highest known for any star.

1984   The Dutch-American Infra-Red Astronomical Satellite (IRAS) detects an excess amount of infrared radiation from Fomalhaut. On the basis of this observation, astronomers conclude that the star must be surrounded by a dusty debris disk.

2008   The visual detection (with the Hubble Space Telescope) of a planet orbiting the bright star Fomalhaut is announced. Fomalhaut b (also known as Dagon) is a Jupiter-like gas giant, probably still surrounded by a huge cloud of accreting gas and dust. It orbits its parent star once every 1,700 years or so on a highly elliptical path that takes it out to a distance of some 28 million miles. Initially, there is some doubt about the true planetary nature of Fomalhaut b, but subsequent observations in 2012 are convincing enough to most astronomers.

▲ Piscis Austrinus appears as a monstrous fish in Johann Bayer's 1603 star atlas *Uranometria*.

◀ NGC 7173 (left), NGC 7174 (center right), and NGC 7176 (lower right) form a small group of interacting galaxies, known as Hickson Compact Group 90.

PUPPIS

PASSPORT

Latin name: Puppis	**Area:** 673.4 square degrees
English name: Stern	**Number of naked-eye stars:** 237
Genitive: Puppis	**Bordering constellations:** Monoceros, Canis Major, Columba, Pictor, Carina, Vela, Pyxis, Hydra
Abbreviation: Pup	
Origin: de Lacaille	**Best visibility:** December–February, south of 35° north

▶ The Skull and Crossbone Nebula (NGC 2467) contains dozens of newly born protostars.

▼ Globular cluster NGC 2298 may be a former member of the Canis Major dwarf galaxy.

PUPPIS (THE STERN) is a large constellation southeast of Canis Major (the Greater Dog). In ancient times, it used to be part of the even larger constellation Argo Navis (the Ship *Argo*). In 1752, French astronomer Nicolas Louis de Lacaille divided Argo Navis into three smaller constellations: Puppis, Vela (the Sails), and Carina (the Keel).

Puppis is crossed by the central band of the Milky Way, and contains a large number of nebulae, star clusters, and bright stars. The brightest star in the constellation is Naos (Zeta Puppis, or ζ Pup)—an extremely hot supergiant at some 1,100 light-years from Earth.

▲ The beautiful open cluster M47 in Puppis contains just a few dozen young, blue stars.

TIMELINE

1654   In Italy, Giovanni Batista Hodierna is the first to observe the open star cluster in Puppis that is now known as M47. At a distance of some 1,600 light-years from Earth, it contains just a few dozen stars in a region that measures 12 light-years across.

1790   English astronomer William Herschel discovers NGC 2440, a planetary nebula in Puppis with a very hot white dwarf star at its center. The star has a surface temperature of some 400,000 degrees F and produces at least a thousand times more energy than our sun, despite its small size.

1826   James Dunlop discovers NGC 2298, a globular cluster in Puppis at a distance of approximately 30,000 light-years.

▲ Apart from three Neptunus-like planets, the star HD69830 is also orbited by a belt of asteroids.

◄ In the core of planetary nebula NGC 2440 is one of the hottest white dwarf stars known.

2006   At the La Silla Observatory in Chile, three Neptune-like gaseous planets are found orbiting the star HD 69830. The star is slightly smaller and less massive than our sun. It is 40 light-years away. The outermost planet is in the star's habitable zone.

2009   NASA's Spitzer Space Telescope detects forty-five protostars in NGC 2467, an active region of star formation in Puppis also known as the Skull and Crossbones Nebula.

PYXIS

PASSPORT

Latin name: Pyxis	**Area:** 220.8 square degrees
English name: Mariner's Compass	**Number of naked-eye stars:** 41
Genitive: Pyxidis	**Bordering constellations:** Hydra, Puppis, Vela, Antlia
Abbreviation: Pyx	**Best visibility:** January–February, south of 50° north
Origin: de Lacaille	

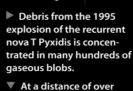

▷ **Debris from the 1995 explosion of the recurrent nova T Pyxidis is concentrated in many hundreds of gaseous blobs.**

▽ **At a distance of over 10,000 light-years in Pyxis, NGC 2818 is one of the few planetary nebulae in an open star cluster.**

PYXIS (THE MARINER'S COMPASS) is a small constellation in the southern sky, south of the star Alphard in Hydra (the Sea Serpent). Despite its nautical name and its proximity to Puppis (the Stern), Vela (the Sails), and Carina (the Keel), it never was part of the ancient constellation Argo Navis (Ship *Argo*).

◁ **NGC 2613 is one of just a handful of galaxies in Pyxis. A spiral like our own Milky Way, it is some 70 million light-years away.**

TIMELINE

1763 Nicolas Louis de Lacaille introduces Pyxis as one of fourteen new constellations in the southern sky. Most of them are named after scientific or industrial instruments.

1995 The Hubble Space Telescope carries out detailed observations of the recurrent nova T Pyxidis, which explodes every twenty years or so on average.

2007 A massive Jupiter-like planet is discovered in orbit around the red dwarf star Gliese 317, which is some 50 light-years away in Pyxis. The system probably also contains a second gas giant.

RETICULUM

PASSPORT

Latin name: Reticulum	**Area:** 113.9 square degrees
English name: Reticle	**Number of naked-eye stars:** 23
Genitive: Reticuli	**Bordering constellations:** Horologium, Hydrus, Dorado
Abbreviation: Ret	**Best visibility:** October–December, south of 20° north
Origin: de Lacaille	

▼ **According to Betty and Barney Hill, their alien abductors came from the binary star Zeta Reticuli (right). The star at left is Kappa Reticuli.**

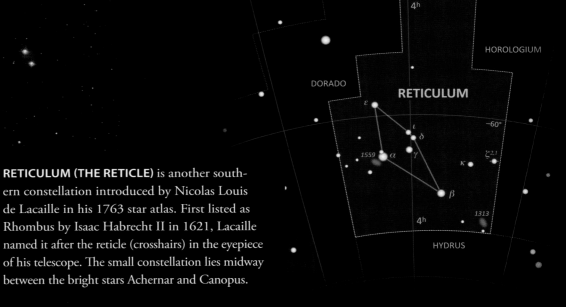

RETICULUM (THE RETICLE) is another southern constellation introduced by Nicolas Louis de Lacaille in his 1763 star atlas. First listed as Rhombus by Isaac Habrecht II in 1621, Lacaille named it after the reticle (crosshairs) in the eyepiece of his telescope. The small constellation lies midway between the bright stars Achernar and Canopus.

TIMELINE

1826   Scottish astronomer James Dunlop discovers the barred spiral galaxy NGC 1313 in Reticulum. Nicknamed the Topsy-Turvy galaxy, it has a very high rate of star formation.

1961   The New Hampshire couple Betty and Barney Hill claim to have been abducted for two days by aliens from a planet orbiting the star Zeta Reticuli, or ζ Ret. It is a binary star at a distance of 39 light-years. So far, no planets have been discovered orbiting either companion.

2005   A supernova (SN 2005df) is detected in the small spiral galaxy NGC 1559 in Reticulum, some 50 million light-years away.

▲ **Supernova SN 2005df can be seen just above the compact galaxy NGC 1559.**

▶ **Bursts of star formation adorn the spiral arms of NGC 1313.**

CHANGING VIEWS

THE APPEARANCE OF THE NIGHT SKY changes throughout the night, because of Earth's rotation. Just after sunset, you may see Orion high up in the sky while Pisces (the Fishes) is low above the eastern horizon. But five hours later, Orion is setting in the west, Pisces reaches its highest point above the horizon, and Aquila (the Eagle) appears in the east.

The appearance of the night sky also changes over the course of a year. The Earth orbits the sun, and when we are looking away from the sun to enjoy the night sky, we see different constellations in March than we do in September. As a result, each season has its own characteristic constellations. For instance, Orion and Gemini (the Twins) are invisible in June, because the sun is in that part of the sky around that time. For the same reason, Scorpius (the Scorpion) and Sagittarius (the Archer) cannot be seen in December.

But there's a third change in the appearance of the night sky, albeit very slow. Thousands of years ago, the sky looked different than it does today, as shown in the illustration on this page. Both views show the southern sky for an observer at a geographical latitude of 40 degrees north (New York or Madrid) in mid-May around 11:00 p.m. local time. But the upper view is for the early twenty-first century, while the lower is drawn for 4000 BC.

While the relative positions of the stars don't change too much over a period of six thousand years, the orientation of the Earth's axis in space *does* change over the millennia. As a result, the appearance of the sky for a particular date and time was quite different in the past than it is now. For instance, stars and constellations that cannot be seen from 40° north today, like Rigil Kentaurus

(Alpha Centauri, or α Cen) and Crux (the Southern Cross), were easily visible from this latitude six thousand years ago.

This slow but steady shift is known as precession. Another result of precession is that our current Pole Star, Polaris (Alpha Ursae Minoris, or α UMi), no longer will serve this role in the distant future. Around 3000 BC, the Earth's axis pointed in the direction of the star Thuban (Alpha Draconis, or α Dra); some 5,500 years in the future, Alderamin (Alpha Cephei, or α Cep) will be our Pole Star.

SOUTH

SOUTH

Top: Looking south in May at approximately 11:00 p.m., today.

Bottom: Looking south around the same time, but in 4000 BC.

SAGITTA

DELPHINUS

VULPECULA

△ PSR B1957+20

● FG

+20°

η

γ

M71

δ

α
● Sham

β

19h

HERCULE

SAGITTA

+ WR124

20h

AQUILA

PASSPORT

Latin name: Sagitta	**Area:** 79.9 square degrees
English name: Arrow	**Number of naked-eye stars:** 26
Genitive: Sagittae	**Bordering constellations:** Vulpecula, Hercules, Aquila, Delphinus
Abbreviation: Sge	
Origin: Ptolemy	**Best visibility:** June–August, north of 65° south

TIMELINE

1746   Swiss astronomer Jean-Philippe de Chéseaux discovers the very loose globular cluster M71, at 13,000 light-years away in Sagitta.

1938   Observations by Paul Merrill reveal that the young, massive star WR 124 in Sagitta is a fast "runaway star," speeding through our Milky Way galaxy at 125 miles per second.

1943   Astronomers discover the peculiar variable star FG Sagittae. Located in a small planetary nebula, the star has markedly changed its color over the decades.

1988   A rapidly rotating millisecond pulsar (PSR B1957+20) is discovered in Sagitta, orbited by a low-mass brown dwarf. The dwarf star is eroded away by the fierce radiation of the pulsar, aptly nicknamed the Black Widow pulsar.

SAGITTA (THE ARROW) is a very small but distinctive constellation just north of the bright star Altair. It lies in the Milky Way, and contains quite a few planetary nebulae and star clusters.

▶ The young, massive star WR124 blows its outer layers into space while careening through the Milky Way at a few hundred kilometers per second.

▼ This X-ray image reveals the bow shock of the Black Widow pulsar PSR B1957+20, which moves through our galaxy at almost a million kilometers per hour.

▲ M71 is one of the smallest globular clusters known. For a long time, it was thought to be an open star cluster.

SAGITTARIUS

PASSPORT

Latin name: Sagittarius	**Area:** 867.4 square degrees
English name: Archer	**Number of naked-eye stars:** 194
Genitive: Sagittarii	**Bordering constellations:** Scutum, Serpens Cauda, Ophiuchus, Scorpius, Corona Australis, Telescopium, Indus (corner), Microscopium, Capricornus, Aquila
Abbreviation: Sgr	
Origin: Ptolemy	
	Best visibility: June–July, south of 40° north

AQUILA

20ʰ   19ʰ

6818
6822

SCUTUM   SERPENS CAUDA

                                    M17 Omega Nebula
                                         18ʰ
ρ¹   M18
                    6716   γ   M24
            π   M25   Star Cloud   M23
Albaldah   –20°   SGR 1806-20   –20°
        o   μ
ECLIPTIC   4   2
–20°   M21
1   M22   M20 Trifid Nebula
M75   Kaus Borealis M28   December
                         σ   Nunki   λ   M8   Solstice
CAPRICORNUS   Lagoon Nebula
                    Wow! signal   φ
                         τ   x
                    3   Kaus Media   Quintuplet Clust.
                    ξ   δ   Galactic Center
                    Ascella   M54
–30°   M55   Sag DEG   γ
                    RR   Alnasl
                         RY   M70   M69
                    SAGITTARIUS
                                         SCORPIUS
                         ε   Kaus Australis
ϑ   η
                    19ʰ   6723   18ʰ
ROSCOPIUM

                    CORONA AUSTRALIS

–40°   RT
ι
                    α
                    Rukbat

                    β¹
                    Arkab Prior
20ʰ   β²   Arkab Posterior
                                         ARA
        TELESCOPIUM

▶ **The Trifid Nebula is a glowing stellar nursery, criss-crossed by dark lanes of dust.**

▶ **The compact Quintuplet Cluster can only be seen at infrared wavelengths: Most of the cluster's visible light is absorbed by galactic dust.**

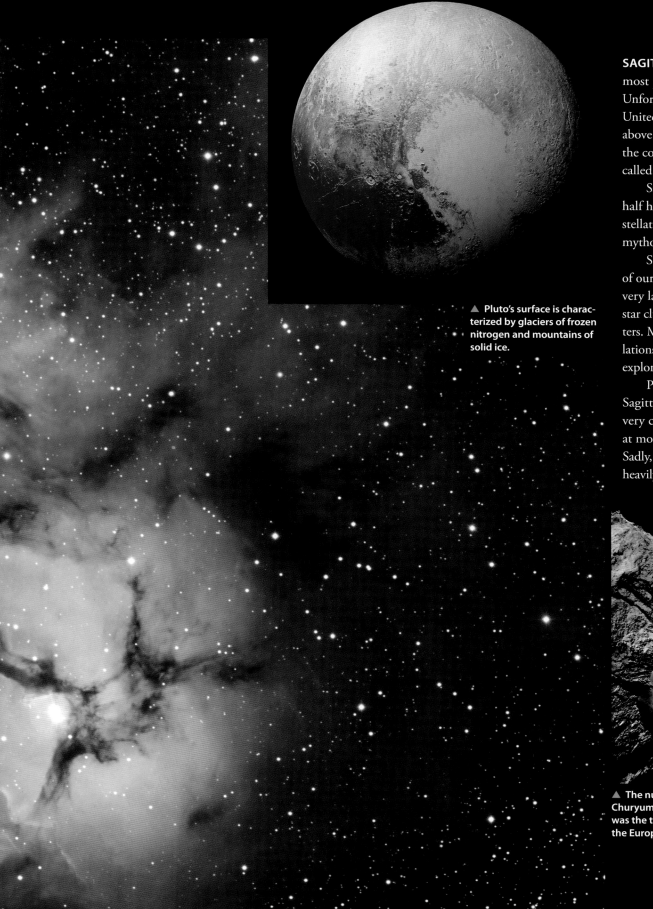

SAGITTARIUS (THE ARCHER) is one of the most prominent constellations of the zodiac. Unfortunately, for observers in large parts of the United States and Europe, it never rises very high above the southern horizon. The central part of the constellation forms a distinctive asterism, called the Teapot.

Sagittarius is depicted as a centaur (half man half horse), wielding bow and arrows. The constellation has also been identified with the Greek mythological satyr-like creature Crotus, son of Pan.

Sagittarius lies in the direction of the center of our Milky Way galaxy. As a result, it contains a very large number of star-forming regions, open star clusters, planetary nebulae, and globular clusters. Moreover, since it is one of the twelve constellations of the zodiac, many highlights in planetary exploration occurred within its boundaries.

Probably the most exciting object in Sagittarius is the supermassive black hole in the very core of our Milky Way galaxy, weighing in at more than 4 million times the mass of the sun. Sadly, the bright and energetic Galactic Center is heavily obscured by thick clouds of absorbing dust.

▲ Pluto's surface is characterized by glaciers of frozen nitrogen and mountains of solid ice.

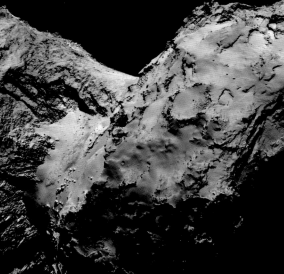

▲ The nucleus of comet 67P/ Churyumov-Gerasimenko was the target destination of the European Philae lander.

▶ The Hubble Space Telescope captured this dramatic image of the Lagoon Nebula on the occasion of its twenty-eighth birthday in April 2018.

▼ The star cluster Palomar 12 in Sagittarius used to be part of the Sagittarius Dwarf Elliptical galaxy (SagDEG).

◀ The first-ever close-up image of a cometary nucleus was obtained in 1986, when the Giotto spacecraft encountered Halley's Comet in Sagittarius.

▶ Neptune, as imaged by NASA's Voyager 2 spacecraft.

1654 Italian astronomer Giovanni Battista Hodiema discovers the Lagoon Nebula (M8) in Sagittarius—one of the constellation's three bright emission nebulae. (The other two are the Omega Nebula [M17] and the Trifid Nebula [M20].) The Lagoon Nebula is a giant stellar nursery at a distance of approximately 4,000 light-years.

1931 American physicist and radio engineer Karl Jansky is the first to detect cosmic radio waves coming from the direction of the Milky Way's center in Sagittarius. The point-like radio source, coinciding with our galaxy's supermassive black hole, is now called Sagittarius A*.

1977 Using Ohio State University's Big Ear radio telescope, astronomer Jerry Ehman discovers what looks like a brief, artificial transmission from an alien civilization, coming from the direction of Sagittarius. It became known as the Wow! signal; its true nature and origin is still unknown.

1983 Using infrared detectors to peer through the Milky Way's dust, astronomers discover what later turns out to be a very rich and compact star cluster close to the Galactic Center. The Quintuplet Cluster contains the Pistol Star, one of the most massive and luminous stars known in our galaxy. A few years later, the neighboring and even more compact Arches Cluster is discovered.

1986 ❶ Comet Halley is visible in Sagittarius when the European Giotto spacecraft makes its spectacular flyby at just a few hundred miles from Halley's nucleus—the first-ever close-up study of a comet.

1989 ❷ The dishes of NASA's Deep Space Network are aimed at Sagittarius to receive the first-ever close-up images of Neptune, the outermost planet in the solar system, when Voyager 2 makes its historic flyby in August.

1994 Behind the Milky Way's center, astronomers detect a dwarf elliptical galaxy (Sag DEG) that contains four globular clusters. It is in the process of being tidally disrupted by the Milky Way's gravity. Eventually, all of its stars will become part of the Milky Way.

2004 An extremely powerful burst of gamma rays, produced by the highly magnetized neutron star SGR 1806-20, is so energetic that it has a measurable effect on Earth's ionosphere, despite the fact that it originated at a distance of some 50,000 light-years.

2014 ❸ The small Philae lander, part of the European Rosetta spacecraft, touches down on the surface of comet 67P/Churyumov-Gerasimenko, while the comet is traversing Sagittarius, way beyond the orbit of Jupiter.

2015 ❹ In July, NASA's New Horizons spacecraft, launched in 2006, makes its dramatic flyby of Pluto and its large moon Charon while the remote dwarf planet is slowly traversing the Milky Way in Sagittarius.

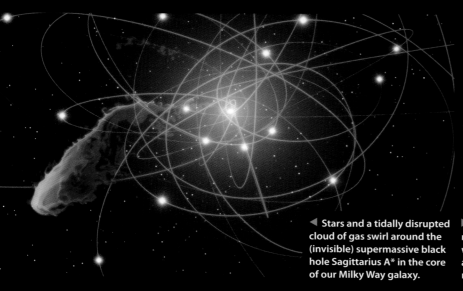

◀ **Stars and a tidally disrupted cloud of gas swirl around the (invisible) supermassive black hole Sagittarius A* in the core of our Milky Way galaxy.**

▶ **The powerful burst of radiation from SGR 1806-20 was probably triggered by a rupture in the magnetic neutron star's crust.**

SCORPIUS

PASSPORT

Latin name: Scorpius		**Area:** 496.8 square degrees	
English name: Scorpion		**Number of naked-eye stars:** 167	
Genitive: Scorpii		**Bordering constellations:** Libra, Lupus, Norma, Ara, Corona Australis, Sagittarius, Ophiuchus	
Abbreviation: Sco			
Origin: Ptolemy		**Best visibility:** May–June, south of 40° north	

SCORPIUS (THE SCORPION) is one of the most conspicuous constellations of the zodiac. It has a very distinctive shape that resembles a real scorpion. The brightest star in the constellation, Antares (Alpha Scorpii, or α Sco) marks the animal's heart. The name *Antares* translates as "rival to Mars," referring to the ruddy color of the star. In classical times, the stars of the neighboring constellation Libra (the Scales) represented the claws of the celestial scorpion.

Like Sagittarius (the Archer), Scorpius lies partly in the central band of the Milky Way and contains a large number of nebulae and star clusters. The constellation is also home to the first X-ray source that was found outside our solar system.

Although Scorpius is part of the zodiac, the sun, moon, and planets spend only a little time within its boundaries.

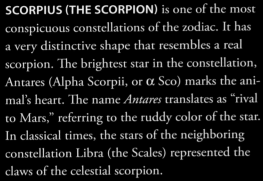

▲ This artist's impression shows a nova outburst of a white dwarf in a binary system. Nova Sco 1437 must have been a similar event.

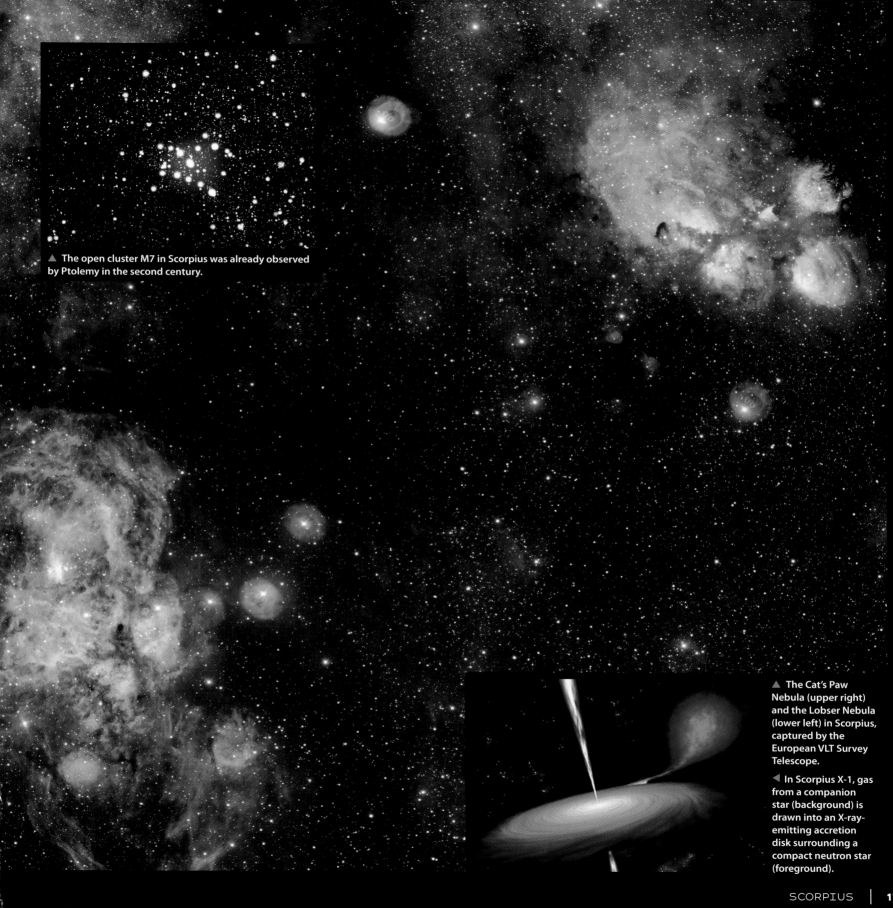

▲ The open cluster M7 in Scorpius was already observed by Ptolemy in the second century.

▲ The Cat's Paw Nebula (upper right) and the Lobser Nebula (lower left) in Scorpius, captured by the European VLT Survey Telescope.

◀ In Scorpius X-1, gas from a companion star (background) is drawn into an X-ray-emitting accretion disk surrounding a compact neutron star (foreground).

◀ This image of the surface of Antares, the brightest star in Scorpius, is the most detailed image ever obtained of any star other than our own sun.

▼ M80 is one of the densest globular clusters in our Milky Way galaxy.

◀ The Bug Nebula (NGC 6302) is a bipolar planetary nebula in Scorpius, at a distance of some 4,000 light-years.

▲ Thousands of individual stars can be seen on this detailed image of the core of the globular cluster M4.

30 The great Greek astronomer Claudius Ptolemy is the first to describe a patch of nebulosity in Scorpius. We now know it is actually a bright open star cluster. Its official designation is M7, but it's often called Ptolemy's Cluster.

437 Korean astronomers observe a bright nova in Scorpius—an energetic explosion on the surface of a white dwarf star in a binary system. Today, an expanding shell of gas can be seen at the site of the stellar outburst. The white dwarf is still producing smaller explosions on a semiregular basis.

746 Swiss astronomer Jean-Philippe de Chéseaux discovers M4, a bright globular star cluster. On the sky, M4 is close to the bright star Antares, but in reality it is much farther away—at least some 7,000 light-years.

781 Another globular cluster, M80, is found by Charles Messier. At a distance of more than 30,000 light-years, it is one of the densest star clusters known. It contains a large number of so-called blue stragglers—seemingly young, hot stars that are the results of stellar mergers.

962 Using a sounding rocket, a team led by Italian physicist Riccardo Giacconi discovers Scorpius X-1, the first X-ray source outside our solar system. It consists of a compact neutron star that devours matter from an orbiting companion.

1994 A nova outburst is observed in the southern part of Scorpius. Nova Sco 1994 emits large amounts of high-energy gamma rays, observed by NASA's Compton Gamma-Ray Observatory, and is also catalogued as GRO J1655-40. The "microquasar" consists of a black hole seven times more massive than the sun, which is orbited by a normal star once every 2.6 days.

2000 Within just a few weeks, the brightness of the star Dschubba (Delta Scorpii, or δ Sco) increases by 50 percent. Since then, it has remained variable, sometimes becoming the second-brightest star in the constellation.

2000 The presence of a giant planet in a wide orbit around pulsar PSR B1620-26 is confirmed. The pulsar resides in the outskirts of globular cluster M4, and forms a close binary pair with a white dwarf star. The planet is about 2.5 times as massive as Jupiter and may be one of the oldest exoplanets found.

2003 Astrobiologist Margaret Turnbull points out that the star 18 Scorpii is one of the most promising nearby stars to host extraterrestrial life on one of its planets. 18 Sco is a "solar analog"—almost an identical twin of our sun. However, no planets have yet been discovered orbiting the star.

2017 Using the Very Large Telescope Interferometer to study Antares in as much detail as possible, European astronomers capture the best-ever image of the surface of another star. Antares is a red supergiant star at a distance of approximately 550 light-years.

◀ **Giant planet PSR 1620-46b (foreground) is orbiting a tight binary, consisting of a pulsar and a white dwarf.**

SCULPTOR

PASSPORT

Latin name: Sculptor	**Area:** 474.8 square degrees
English name: Sculptor	**Number of naked-eye stars:** 52
Genitive: Sculptoris	**Bordering constellations:** Aquarius, Piscis Austrinus, Grus, Phoenix, Fornax, Cetus
Abbreviation: Scl	
Origin: de Lacaille	**Best visibility:** September–October, south of 50° north

SCULPTOR—ORIGINALLY NAMED Apparatus Sculptoris (the Sculptor's Studio) by Nicolas Louis de Lacaille—is a faint constellation south of Cetus (the Whale) and east of Piscis Austrinus (the Southern Fish). It is hard to recognize, as it contains no bright stars and lacks a distinctive shape.

Sculptor is home to the south galactic pole, which means that the constellation provides an unobstructed view of the distant universe. As a result, many galaxies and galaxy clusters can be found within its boundaries.

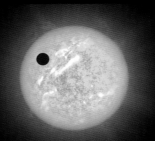

◄ **Exoplanet WASP-8b,** discovered with the transit technique, orbits its host star in the "wrong" direction.

► **The Sculptor Dwarf was the first small Milky Way companion to be discovered.**

TIMELINE

1783 Caroline Herschel, sister of William, discovers the Sculptor galaxy (NGC 253), a beautiful barred spiral galaxy at some 12 million light-years away that displays a high star-forming activity. NGC 253 has been nicknamed the Silver Dollar galaxy.

1937 The Sculptor Dwarf galaxy is discovered by American astronomer Harlow Shapley. It is the first dwarf companion to our Milky Way to be found.

1941 American-Swiss astronomer Fritz Zwicky discovers the Cartwheel galaxy, at some 500 million light-years away in Sculptor. The ring and "spokes" around the galaxy were produced by a head-on galactic collision some 200 million years ago.

2008 The Wide Angle Search for Planets (WASP) detects a Jupiter-like planet in Sculptor orbiting its host star (WASP-8) in a retrograde orbit—that is, opposite to the star's rotation.

2014 The Hubble Space Telescope obtains the deepest-ever image of a distant cluster of galaxies, Abell 2744, in Sculptor, at a distance of almost 4 billion light-years. Also known as Pandora's Cluster, it is actually a smash-up of four smaller galaxy clusters.

2016 While testing a new camera, Argentinian amateur astronomer Victor Buso witnesses the very first brightening of a supernova in the galaxy NGC 613, in the eastern part of Sculptor. SN 2016gkg is the first supernova for which this "shock breakout" has been observed.

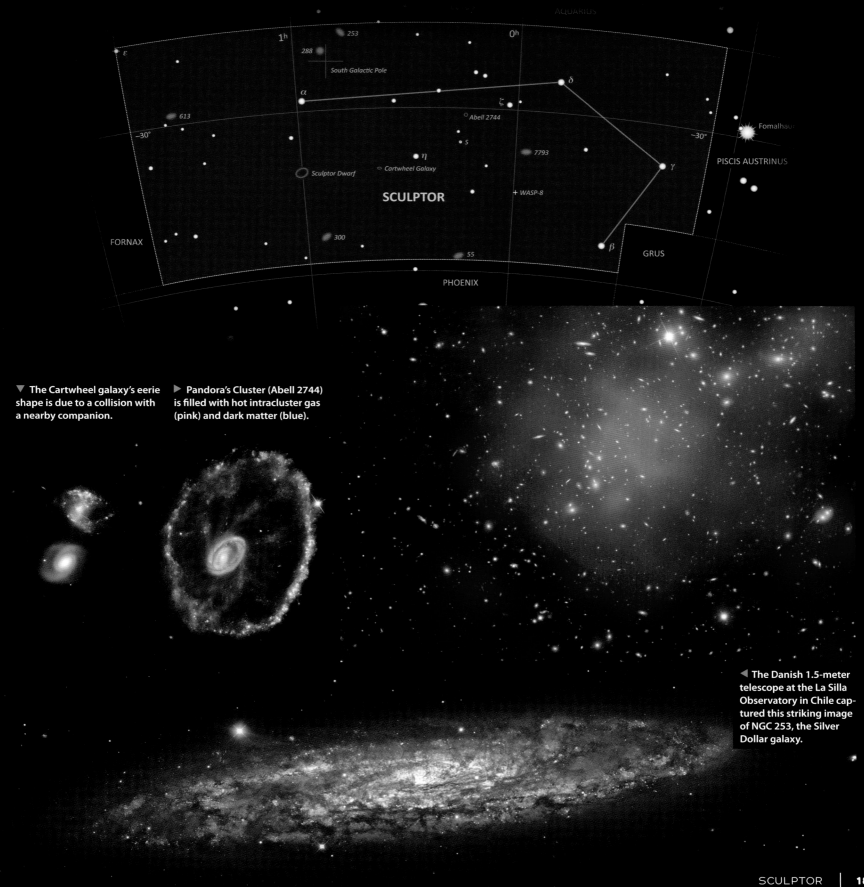

1ʰ

253

288

South Galactic Pole

0ʰ

AQUARIUS

δ

ε

α

ζ

Abell 2744

613

−30°

S

7793

−30°

Fomalhaut

PISCIS AUSTRINUS

η

Sculptor Dwarf

Cartwheel Galaxy

γ

+ WASP-8

SCULPTOR

FORNAX

300

55

β

GRUS

PHOENIX

▼ **The Cartwheel galaxy's eerie shape is due to a collision with a nearby companion.**

▶ **Pandora's Cluster (Abell 2744) is filled with hot intracluster gas (pink) and dark matter (blue).**

◀ **The Danish 1.5-meter telescope at the La Silla Observatory in Chile captured this striking image of NGC 253, the Silver Dollar galaxy.**

SCUTUM

PASSPORT

Latin name: Scutum	**Area:** 109.1 square degrees
English name: Shield	**Number of naked-eye stars:** 29
Genitive: Scuti	**Bordering constellations:** Serpens Cauda, Sagittarius, Aquila
Abbreviation: Sct	**Best visibility:** June–July, south of 70° north
Origin: Hevelius	

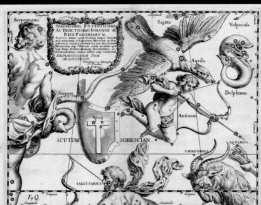

SCUTUM (THE SHIELD) is a small and inconspicuous constellation between Aquila (the Eagle) and Sagittarius (the Archer). It was introduced by Polish astronomer Johannes Hevelius as Scutum Sobiescianum, after the heraldic shield of Polish king Jan III Sobieski, who won the 1683 Battle of Vienna.

Scutum lies in the central band of the Milky Way, and contains many star clusters, nebulae, and luminous star clouds. It also marks the direction in which NASA's space probe Pioneer 11 is leaving the solar system.

▲ **Scutum Sobiescianum as it appears in the 1687 star atlas of Johannes Hevelius.**

▶ **Infrared image of star clusters, nebulae, and dust clouds in Scutum, obtained by NASA's Spitzer Space Telescope.**

▶ **The Hubble Space Telescope captured this image of the core of the Wild Duck Cluster (M11).**

▼ Artistic rendering of Pioneer 11's encounter with Saturn and its moons (in 1979). The spacecraft is now flying in the direction of Scutum.

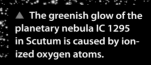

▲ The greenish glow of the planetary nebula IC 1295 in Scutum is caused by ionized oxygen atoms.

TIMELINE

1681   German astronomer Gottfried Kirch discovers a very rich open star cluster in the part of the sky that now belongs to Scutum. In 1764, Charles Messier incorporates the cluster in his famous catalog of nebulous objects as M11. Because of its shape, resembling a flying flock of ducks, M11 is also known as the Wild Duck Cluster, a name suggested by William Smyth in 1844. The cluster contains some three thousand stars in a region of just 20 light-years across. It is approximately 220 million years old and some 6,200 light-years away.

1684   Johannes Hevelius first mentions the new constellation Scutum in a seventeenth-century scientific journal, before incorporating it on the most lavishly illustrated page of his 1687 star atlas *Firmamentum Sobiescianum*.

1900   At Lick Observatory in California, astronomers discover that the variable star Delta Scuti, or δ Sct, is actually pulsating. The star, which brightens and fades every 4.65 hours, is now the prototype of a class of variable stars known as dwarf Cepheids.

1973   ❶ Pioneer 11 is launched as the first spacecraft to visit the giant planet Saturn (in 1979), after visiting Jupiter, like its sister ship Pioneer 10 did. It is now leaving the solar system in the direction of Scutum. However, it will be tens of thousands of years before it encounters another star.

SERPENS

SERPENS (THE SERPENT) is a constellation that actually consists of two parts: Serpens Caput (the Serpent's Head) and Serpens Cauda (the Serpent's Tail). It was already listed by Greek astronomer Ptolemy as the snakelike creature held by Ophiuchus (the Serpent Bearer), who represents the healer Asclepius. But when Belgian astronomer Eugène Delporte established the modern borders of the constellations in 1930, he had to split the sinuous creature in two halves.

The head of Serpens is easy to recognize as a small triangle of stars, north of the bright star Alpha Serpentis, or α Ser, which is also known as Unukalhai, from the Arabic *Unuq al-Hayyah* (Neck of the Serpent). The Serpent's Tail is smaller and less conspicuous.

By far the most famous celestial object in the constellation is the Eagle Nebula, with its dramatic Pillars of Creation—three towering, star-spawning dust clouds that were imaged by the Hubble Space Telescope in 1995.

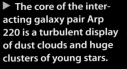

▶ **The core of the interacting galaxy pair Arp 220 is a turbulent display of dust clouds and huge clusters of young stars.**

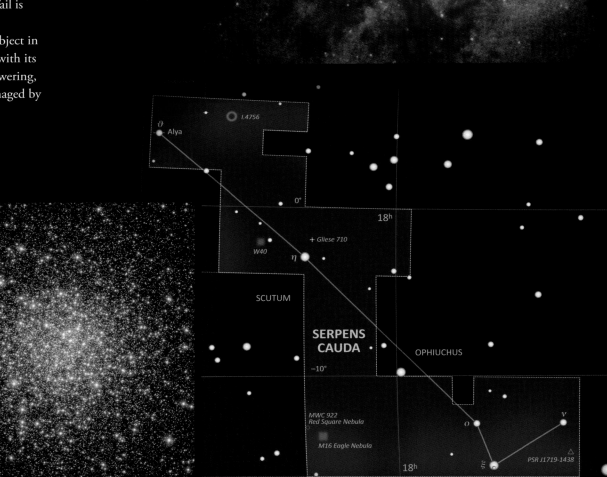

▶ **M5 is a bright globular cluster in Serpens, containing over a million old stars.**

CORONA
BOREALIS

16ʰ

◇ Arp 220

Hoag's Object ◇

Seyfert's
Sextet ◇

+20°

BOÖTES

κ

γ   β
R

△ PSR B1534+12

δ

+10°

λ

α Unukalhai

HERCULES

SERPENS CAPUT

VIRGO

M5

0°

OPHIUCHUS

16ʰ

μ

LIBRA

SCORPIUS

▲ The famous Pillars of
Creation can be seen in the
center of this wide-angle
image of the Eagle Nebula.

▶ Right now, it's still a
dim star at 64 light-years'
distance, but in 1.3 million
years, Gliese 710 (center) will
approach our solar system
to within 0.3 light-years,
becoming one of the brightest stars in the sky.

▶ MWC 922, also known as
the Red Square Nebula, is one
of the most symmetrical planetary nebulae known. It is at
5,000 light-years' distance in
Serpens Cauda.

1702 German astronomer Gottfried Kirch discovers the bright globular cluster M5 in Serpens Caput. At a distance of some 25,000 light-years, this collection of hundreds of millions of stars can just be seen with the naked eye under perfect observing conditions.

1745 The Eagle Nebula (M16) and its associated star cluster are first described by Swiss astronomer Jean-Philippe de Chéseaux. It is an active site of star formation at a distance of 7,000 light-years in Serpens Cauda, actually very close to the Omega Nebula (M17) in Sagittarius (the Archer).

1950 Arthur Hoag discovers a remarkable galaxy, now known as Hoag's Object, in Serpens Caput. It is a so-called ring galaxy, with a core of old stars and a circular ring of much younger stars. It may have been a barred spiral galaxy whose spiral arms got detached somehow.

1951 While using photographic plates made at the Barnard Observatory of Vanderbilt University, American astronomer Carl Seyfert comes across a very compact group of six galaxies in Serpens Caput. Four of the galaxies in Seyfert's Sextet are actually very close together in space, their shapes being disturbed by tidal forces. In the future, they will merge into one giant elliptical galaxy.

1966 At the Palomar Observatory in California, Halton Arp publishes his *Atlas of Peculiar Galaxies*, containing 338 examples of interacting or otherwise misshapen galaxies. One of the most famous is Arp 220 in Serpens Caput, the prototype of an ultraluminous infrared galaxy. Arp 220 is actually two galaxies in the process of merging. It has an incredibly high rate of star formation, and frequently produces supernovae.

▶ **The small spiral in Seyfert's Sextet is actually a background object, while the faint "object" at the very right is a tidal tail of one of the four interacting galaxies.**

1999 American radio astronomers find evidence for "geodetic precession" of the spin axis of pulsar PSR B1534+12 in Serpens Caput. The pulsar spins around its axis every 37.9 milliseconds while orbiting another neutron star. The slow wobbling of its spin axis is a relativistic effect, confirming a prediction of Einstein's theory of general relativity.

2010 The European Herschel Space Observatory discovers the prevalence of filamentary structures in stellar nurseries by studying Westerhout 40 (W40), a nearby region of star formation in Serpens Cauda that is partly obscured by dark molecular clouds.

2011 A small, dense companion is found in orbit around the pulsar PSR J1719-1438 in Serpens Cauda. The Jupiter-mass object, nicknamed the Diamond Planet, is probably the eroded remnant of a white dwarf star, consisting almost solely of compressed carbon.

2016 Precision measurements by the European Gaia mission reveal that the dwarf star Gliese 710, at 64 light-years' distance in Serpens Cauda, will approach the sun in about 1.3 million years to a distance of just one-third of a light-year or so. By then, it will be one of the brightest stars in the sky. It may also disturb the solar system's Oort Cloud of comets and produce a shower of cometary impacts on Earth.

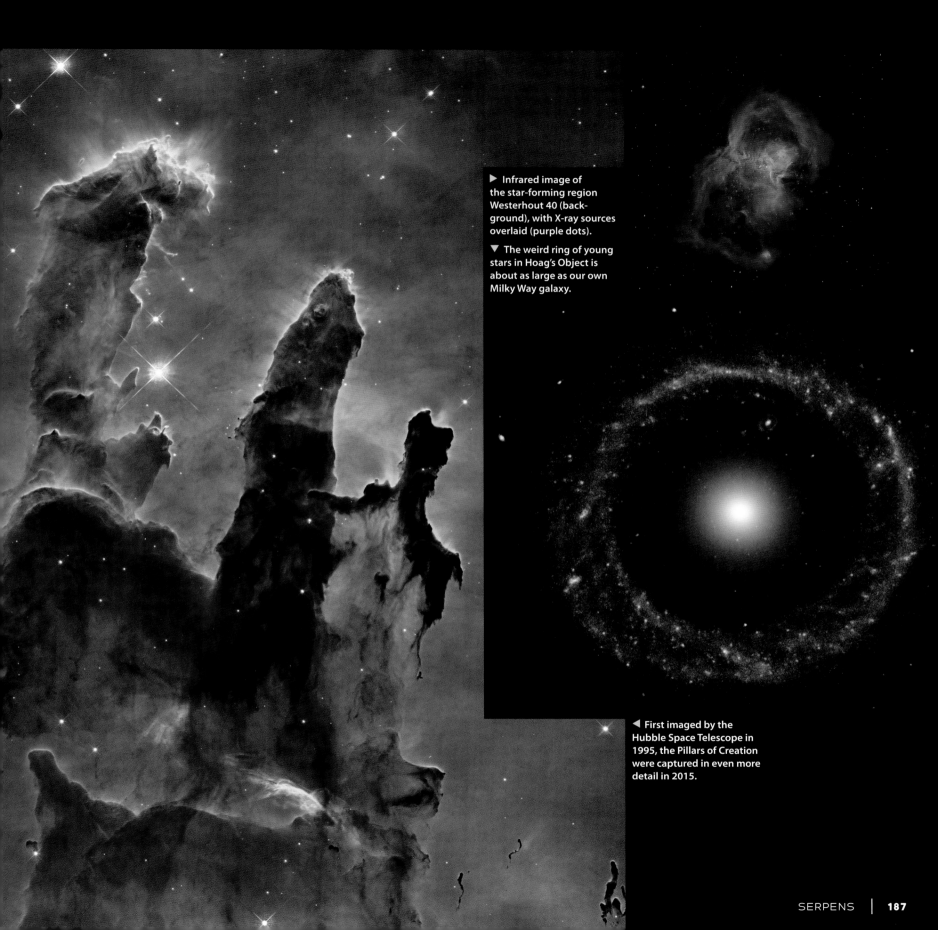

▶ Infrared image of
the star-forming region
Westerhout 40 (back-
ground), with X-ray sources
overlaid (purple dots).

▼ The weird ring of young
stars in Hoag's Object is
about as large as our own
Milky Way galaxy.

◀ First imaged by the
Hubble Space Telescope in
1995, the Pillars of Creation
were captured in even more
detail in 2015.

SEXTANS

PASSPORT

Latin name: Sextans	**Area:** 313.5 square degrees
English name: Sextant	**Number of naked-eye stars:** 38
Genitive: Sextantis	**Bordering constellations:** Leo, Hydra, Crater
Abbreviation: Sex	**Best visibility:** January–March, between 75° north and 80° south
Origin: Hevelius	

SEXTANS (THE SEXTANT) is a nondescript constellation south of Leo (the Lion). The relatively empty part of the sky was named after the famous astronomical and navigational instrument by Polish astronomer Johannes Hevelius, in his 1687 star atlas *Firmamentum Sobiescianum*. The constellation straddles the celestial equator.

Since Sextans is pretty far from the central band of the Milky Way, it provides astronomers with a good view of the distant universe. The constellation contains many galaxies and is also home to the Cosmos Evolution Survey (COSMOS)— the largest survey project ever undertaken by the Hubble Space Telescope.

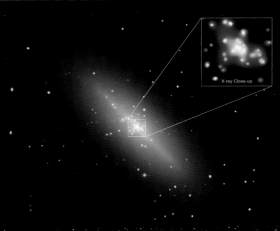

X-ray Close-up

◀ The study of X-ray sources (blue points) in the core of the edge-on galaxy NGC 3115 reveals the existence of a supermassive black hole.

▲ The purple blob marks the X-rays from extremely hot gas in CLJ 1001+0220, the most distant galaxy cluster known.

▶ Deep infrared view of the COSMOS field in Sextans, revealing numerous remote galaxies.

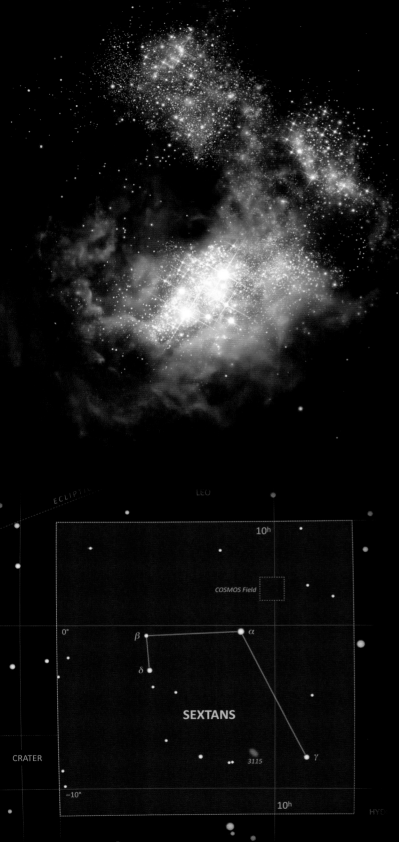

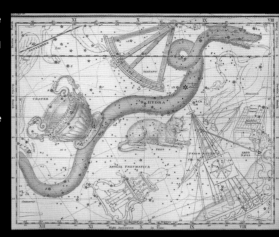

◀ Artist's impression of the bright, star-forming galaxy CR7 as it may have appeared when the universe was less than a billion years old.

▶ Sextans appears in Alexander Jamieson's 1822 star atlas, together with the now obsolete constellation Felis (the Cat).

TIMELINE

1787 English astronomer William Herschel discovers NGC 3115, a nearly edge-on galaxy in Sextans at a distance of just over 30 million light-years. In 2011, X-ray observations reveal that NGC 3115 (sometimes called the Spindle galaxy) harbors a central black hole that is 2 billion times as massive as the sun.

1923 Due to precession—a very slow wobbling motion of Earth's axis—the star Alpha Sextantis, or α Sex, passes the celestial equator in December, moving from the northern to the southern hemisphere of the sky.

2004 Using the Hubble Space Telescope and several large telescopes on the ground, astronomers start COSMOS, a comprehensive survey of a two-square-degree field in Sextans, to study the evolution of galaxies and large-scale structure. Over the years, more than 2 million galaxies are identified and characterized by this survey.

2015 The first convincing evidence for the existence of Population III stars—the very earliest generation of stars in the universe—is found in a surprisingly bright galaxy (nicknamed CR7) at 12.9 billion light-years away in the COSMOS Field, discovered by the European Very Large Telescope.

2016 Also in the COSMOS Field, NASA's Chandra X-ray Observatory, together with instruments on the ground, discovers the most distant cluster of galaxies found to date: CLJ 1001+0220, 11.1 billion light-years away in Sextans.

TAURUS

PASSPORT

Latin name: Taurus

English name: Bull

Genitive: Tauri

Abbreviation: Tau

Origin: Ptolemy

Area: 797.2 square degrees

Number of naked-eye stars: 223

Bordering constellations: Perseus, Aries, Cetus, Eridanus, Orion, Gemini, Auriga

Best visibility: November–December, north of 55° south

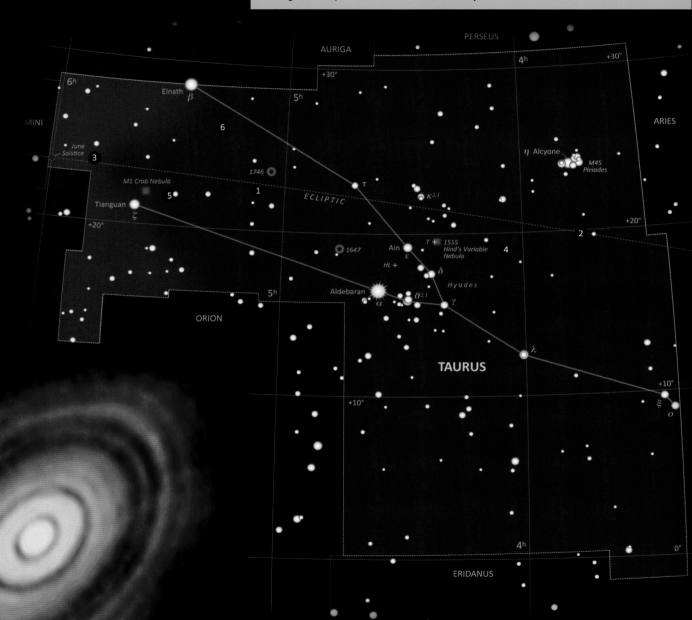

▼ ALMA reveals tell-tale structures in the disk surrounding the baby star HL Tauri, hinting at the formation of planets.

◀ The Pleiades (M45), also known as the Seven Sisters, is the most famous star cluster in the sky.

▼ Multi-wavelength composite of the Crab Nebula, the expanding remnant of Supernova 1054.

TAURUS (THE BULL) is one of the oldest and most recognizable constellations. It really has the appearance of a bull, with the stars El Nath (Beta Tauri, or β Tau) and Zeta Tauri (or ζ Tau) representing the two horns, and the bright orange star Aldebaran (Alpha Tauri, or α Tau) marking the bovine eye. The Pleiades—the most striking star cluster in the night sky, also known as the Seven Sisters—is in the bull's neck. To the naked eye, the cluster looks like a small nebulous cloud of stars.

The constellation is many millennia old—it already appears in an almost twenty-thousand-year-old cave painting in Lascaux, France. To the ancient Greeks, it represented Zeus, who disguised himself as a bull to seduce and kidnap the Phoenician princess Europa. After being carried to the island of Crete, Europa gave birth to Minos, who would become king of Crete and founder of the European civilization.

Taurus is one of the twelve constellations of the zodiac, and it has served as a backdrop for many discoveries and events in solar system science.

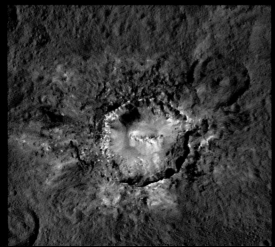

◀ Haulani is a 34-kilometer-wide impact crater on Ceres, the largest asteroid. This image was captured by NASA's Dawn spacecraft.

► The baby star T Tauri, just 1 million years old or so, is the orange star near the center of this image. It is still surrounded by dust-laden clouds of molecular gas.

▼ Numerous haze layers are visible in this Cassini image of Titan's atmosphere.

▼ Io, the innermost of the four major moons of Jupiter, is a weird world of sulphur volcanoes and lava lakes.

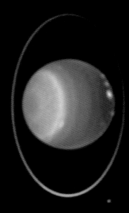

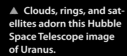

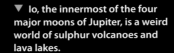

▲ Clouds, rings, and satellites adorn this Hubble Space Telescope image of Uranus.

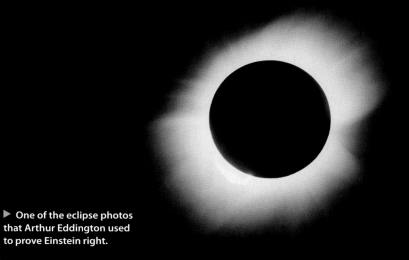

▶ One of the eclipse photos that Arthur Eddington used to prove Einstein right.

◀ Pioneer 10 ventures into deep space, in the direction of Taurus.

TIMELINE

1054   Chinese and Korean court astronomers witness the explosion of a supernova, close to the star Zeta Tauri, or ζ Tau. The remnant of the catastrophic stellar blast—the Crab Nebula, or M1—was discovered in 1731 by John Bevis. It is still expanding at a rate of almost 1,000 miles per second.

1610   ❶ When Jupiter is high in the sky in Taurus, Italian astronomer Galileo Galilei points his homebuilt telescope at the giant planet and discovers its four major moons: Io, Europa, Ganymede, and Callisto. In honor of his discovery, they are collectively known as the Galilean moons.

1690   ❷ John Flamsteed, the first Royal Astronomer, lists a star in Taurus as 34 Tauri without realizing that it is actually a previously unknown planet. Uranus would not be discovered until almost a century later.

1781   ❸ On March 13, from the garden of his house in Bath, England, German-born astronomer William Herschel discovers Uranus, while the distant planet is again slowly crossing Taurus. Apart from a number of comets, Uranus is the first object found to be orbiting the sun that was not already known in antiquity.

1801   ❹ In the early hours of January 1, Giuseppe Piazzi studies a star field in Taurus from the Palermo Observatory on the Italian island of Sicily and discovers a starlike object that moves across the sky. He believes it is yet another planet and calls it Ceres, after the Roman god of fertility. We now know that Ceres is a dwarf planet and the largest member of the asteroid belt between the orbits of Mars and Jupiter.

1852   John Russell Hind discovers the remarkable variable star T Tauri, which illuminates Hind's Variable Nebula (NGC 1555). It becomes the prototype of a class of newly born stars that are still surrounded by the clouds of gas and dust from which they were born. T Tauri is some 460 light-years from Earth.

1919   On May 29, the sun is eclipsed by the moon while it is located close to the Hyades star cluster in Taurus. On the African island of Principe, Arthur Eddington measures the tiny displacement of the Hyades stars, caused by the bending of starlight in the sun's gravitational field—the first confirmation of Einstein's theory of general relativity.

1944   ❺ Dutch-American planetary scientist Gerard Kuiper points his telescope at Taurus to study the large Saturnian moon Titan. Kuiper detects the spectroscopic signature of methane, proving that Titan has an atmosphere of its own.

1972   ❻ NASA launches Pioneer 10, the first spacecraft to visit the outer solar system. After encountering Jupiter in December 1973, the tiny craft is now leaving the solar system in the direction of Taurus.

2014   The Atacama Large Millimeter/submillimeter Array (ALMA) captures a dramatic image of the protoplanetary disk of HL Tauri, a very young T Tauri star at a distance of 450 light-years. Gaps in the disk suggest that planets have already started to form.

TELESCOPIUM

PASSPORT

Latin name: Telescopium	**Area:** 251.5 square degrees
English name: Telescope	**Number of naked-eye stars:** 57
Genitive: Telescopii	**Bordering constellations:** Sagittarius, Corona Australis, Ara, Pavo, Indus, Microscopium (corner)
Abbreviation: Tel	
Origin: de Lacaille	**Best visibility:** June–July, south of 30° north

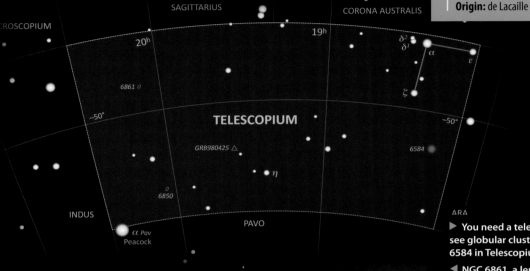

▶ You need a telescope to see globular cluster NGC 6584 in Telescopium.

◀ NGC 6861, a lenticular galaxy with a prominent dust band, is one of the brightest members of the Telescopium Group.

TELESCOPIUM (THE TELESCOPE) is a rather uninspiring galaxy in the southern sky. It was introduced in the eighteenth century by French astronomer Nicolas Louis de Lacaille, who named it after the most important instrument in astronomy.

TIMELINE

1826 Scottish astronomer James Dunlop discovers a globular cluster in Telescopium, now known as NGC 6584. It is some 45,000 light-years away and contains many variable stars.

1836 John Herschel discovers the spiral galaxy NGC 6850, one of the twelve members of the Telescopium Group, at a distance of approximately 120 million light-years.

1998 On April 25, the Dutch-Italian satellite BeppoSAX detects a bright gamma-ray burst (GRB 980425) that turns out to be related to the superluminous supernova SN 1998bw, in a remote galaxy in Telescopium. This is the first evidence that long gamma-ray bursts are related to supernovae.

◀ The arrow points to supernova SN 1998bw in the remote galaxy ESO 184-G82. The supernova also produced a gamma-ray burst.

TRIANGULUM

PASSPORT

Latin name: Triangulum	**Area:** 131.8 square degrees
English name: Triangle	**Number of naked-eye stars:** 25
Genitive: Trianguli	**Bordering constellations:** Andromeda, Pisces, Aries, Perseus
Abbreviation: Tri	**Best visibility:** September–November, north of 50° south
Origin: Ptolemy	

TRIANGULUM (THE TRIANGLE) is a small but quite recognizable constellation between Andromeda and Aries (the Ram). In ancient times, the triangular group of three stars has been associated with both the delta of the river Nile and the Italian island of Sicily.

Triangulum is most famous for harboring the Triangulum galaxy within its borders. Also known as M33, this is the third-largest member of the Local Group of galaxies, after the Andromeda galaxy and our own Milky Way. Despite its distance of 2.3 million light-years and its relatively small size, it can just be seen with the naked eye under extremely favorable observing conditions.

◀ At radio wavelengths (color-coded purple), the spiral arms of M33 reach out much farther from the center than in visible light.

▶ The Triangulum galaxy (M33), as observed by the European VLT Survey Telescope.

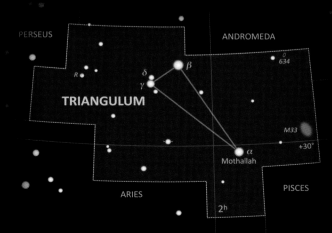

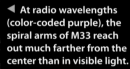

1654   Italian astronomer Giovanni Battista Hodierna is the first to describe a faint smudge of nebulosity, more or less midway between the stars Alpha Arietis, or α Ari, and Beta Andromedae, or β And. Charles Messier makes an independent discovery in 1764 and lists the nebula as the thirty-third object in his famous catalog.

1784   Close to the bright core of M33, William Herschel discovers a smaller nebulous patch, NGC 604. It later turned out to be a giant star-forming region in one of the galaxy's spiral arms.

1923   Dutch astronomer Adriaan van Maanen publishes his (erroneous) observations of rotational motion in M33, which would have put the nebula in our own Milky Way galaxy. Not much later, the extragalactic nature of spiral nebulae was firmly established.

2007   X-ray observations of a source known as M33 X-7 reveal that it is a black hole in a binary system, weighing in at almost sixteen times the mass of the sun—the bulkiest stellar-mass black hole at the time.

2008   A supernova explodes in NGC 634, a rather faint galaxy in Triangulum that was discovered in the nineteenth century by French astronomer Édouard Stephan.

▲ After 30 Doradus in the Large Magellanic Cloud, NGC 604 in M33 is the largest star-forming region in the Local Group of galaxies.

◀ M33 X-7 is a black hole (foreground) orbiting a very massive star. Stellar gas accretes on a disk around the black hole, producing X-rays.

▶ In 2008, the dusty spiral galaxy NGC 634 was home to supernova SN 2008a.

TRIANGULUM AUSTRALE

PASSPORT

Latin name: Triangulum Australe	**Area:** 110.0 square degrees
English name: Southern Triangle	**Number of naked-eye stars:** 35
Genitive: Trianguli Australis	**Bordering constellations:** Norma, Circinus, Apus, Ara
Abbreviation: TrA	**Best visibility:** April–June, south of 15° north
Origin: Keyser and de Houtman	

TRIANGULUM AUSTRALE (THE SOUTHERN TRIANGLE) is a prominent constellation in the southern sky, first described by Dutch sailors Pieter Dirkszoon Keyser and Frederick de Houtman, and officially introduced by Petrus Plancius in 1598.

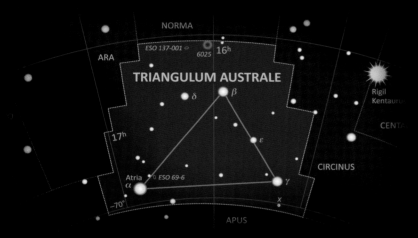

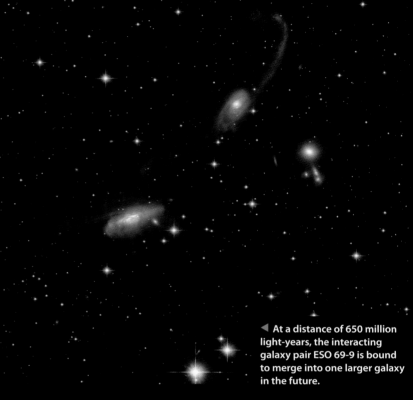

◄ At a distance of 650 million light-years, the interacting galaxy pair ESO 69-9 is bound to merge into one larger galaxy in the future.

▲ The open star cluster NGC 6025 hardly stands out against the background of fainter Milky Way stars.

▶ X-ray observations (color-coded purple) reveal the hot gas that is stripped away from the galaxy ESO 137-001 as it races through intergalactic space.

TIMELINE

1751 While charting the southern sky from Cape Town, French astronomer Nicolas Louis de Lacaille discovers the open star cluster NGC 6025 at the very northern edge of Triangulum Australe. It lies at a distance of 2,700 light-years, in the star-studded central band of the Milky Way.

2008 On the eighteenth anniversary of the Hubble Space Telescope, a large collection of photos of interacting galaxies is published, including the galaxy pair ESO 69-6 in Triangulum Australe.

2014 Observations with the MUSE instrument at the European Very Large Telescope reveal how the galaxy ESO 137-001 is losing much of its gas by ram-pressure stripping as it plunges into the Norma Cluster.

TUCANA

Latin name: Tucana	**Area:** 294.6 square degrees
English name: Toucan	**Number of naked-eye stars:** 45
Genitive: Tucanae	**Bordering constellations:** Phoenix, Grus, Indus, Octans, Hydrus, Eridanus (corner)
Abbreviation: Tuc	
Origin: Keyser and de Houtman	**Best visibility:** April–June, south of 15° north

TUCANA (THE TOUCAN) is a constellation in the southern sky that contains only a few conspicuous stars and lacks a distinctive shape. It is famous, though, for harboring the Small Magellanic Cloud—a nearby satellite galaxy that looks like a detached part of the Milky Way. Like its larger cousin in the constellation Dorado (the Goldfish), it is named after the Portuguese explorer Fernão de Magalhães (Ferdinand Magellan).

Tucana is also home to 47 Tucanae, one of the brightest globular clusters in the sky.

TIMELINE

1598   Flemish astronomer Petrus Plancius introduces the new constellation on a celestial globe, following the suggestion by Dutch sailors Pieter Dirkszoon Keyser and Frederick de Houtman.

1751   Nicolas Louis de Lacaille is the first to describe the globular cluster nature of the nebulous "star" 47 Tucanae, or 47 Tuc. It measures some 120 light-years across and lies at a distance of almost 15,000 light-years.

1908   On the basis of observations of Cepheid variable stars in the Small Magellanic Cloud, American astronomer Henrietta Leavitt published her period-luminosity law (also known as the Leavitt Law), which is a powerful tool in establishing the cosmic distance scale.

1998   After the success of the original Hubble Deep Field observations in Ursa Major (the Great Bear), astronomers carry out a similar campaign in the southern sky, by observing an "empty" patch of sky in Tucana for many hours. The Hubble Deep Field South contains the quasar QSO J2233-606.

2007   An extremely brief but energetic burst of radio waves from the direction of Tucana is discovered in observations made in 2001 by the 209-foot Parkes radio telescope. The Lorimer Burst—named after the leader of the discovery team—is the first fast radio burst (FRB) ever detected; its official designation is FRB 010724.

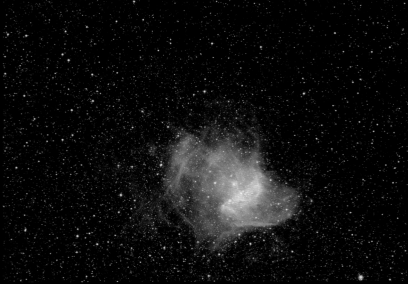

▲ NGC 346 is an active site of star formation in the Small Magellanic Cloud.

◄ Over a thousand faint galaxies, out to a distance of some 12 billion light-years, are visible in the Hubble Deep Field South.

▼ The open star cluster NGC 299 is located in the Small Magelannic Cloud, some 210,000 light-years away.

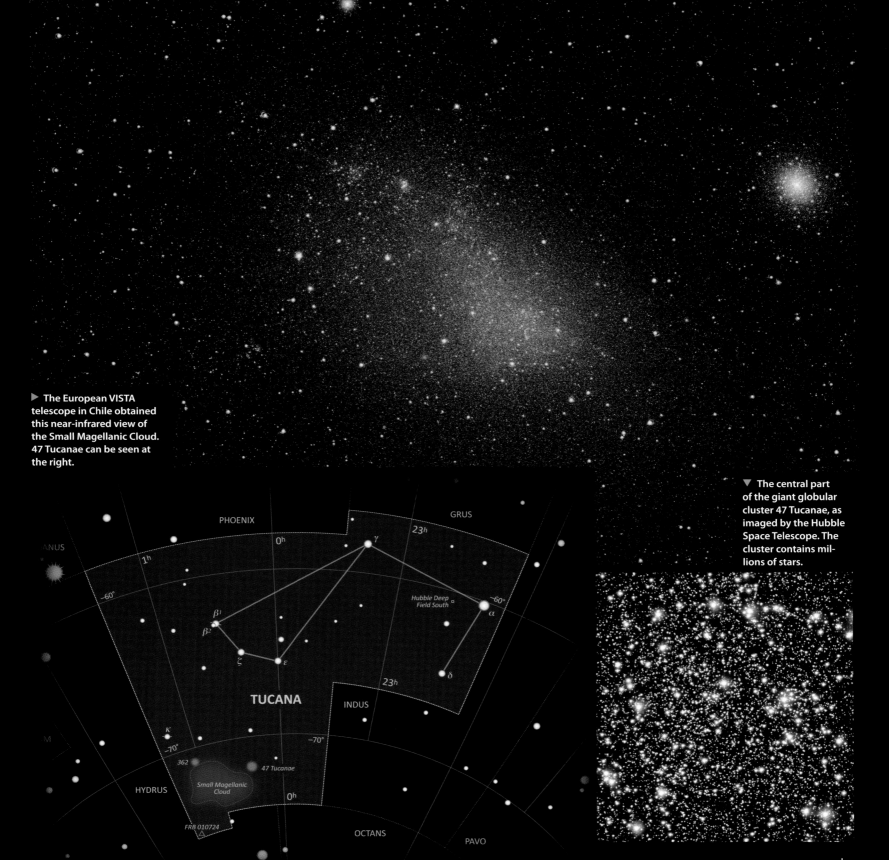

▶ The European VISTA telescope in Chile obtained this near-infrared view of the Small Magellanic Cloud. 47 Tucanae can be seen at the right.

▼ The central part of the giant globular cluster 47 Tucanae, as imaged by the Hubble Space Telescope. The cluster contains millions of stars.

PHOENIX

GRUS

ANUS

0ʰ

1ʰ

23ʰ

γ

–60°

Hubble Deep
Field South □

–60°

β¹

α

β²

ζ

ε

23ʰ

δ

TUCANA

INDUS

κ

M

–70°

–70°

362

47 Tucanae

Small Magellanic
Cloud

HYDRUS

0ʰ

FRB 010724 △

OCTANS

PAVO

URSA MAJOR

PASSPORT

Latin name: Ursa Major	Area: 1,279.7 square degrees
English name: Great Bear	Number of naked-eye stars: 209
Genitive: Ursae Majoris	Bordering constellations: Draco, Camelopardalis, Lynx, Leo Minor, Leo, Coma Berenices, Canes Venatici, Boötes
Abbreviation: UMa	
Origin: Ptolemy	Best visibility: February–April, north of 15° south

URSA MAJOR (THE GREAT BEAR) may well be the most famous constellation in the sky. Its distinctive dipper shape of seven bright stars makes it easy to recognize. Moreover, for observers in most of Europe and North America, the Big Dipper, as the constellation is often called, can be seen all year round.

Ursa Major is also one of the oldest and one of the largest constellations. To the ancients, it represented the nymph Callisto, who had an affair with Zeus. Zeus's wife, Hera, took revenge by turning Callisto into a bear. Callisto's son, the hunter Arcas, was also turned into a bear (Ursa Minor) by Zeus to prevent him accidentally killing his mother.

Ursa Major is well removed from the band of our own Milky Way galaxy, and it presents astronomers with an unobstructed view of the distant universe. It also contains the most famous

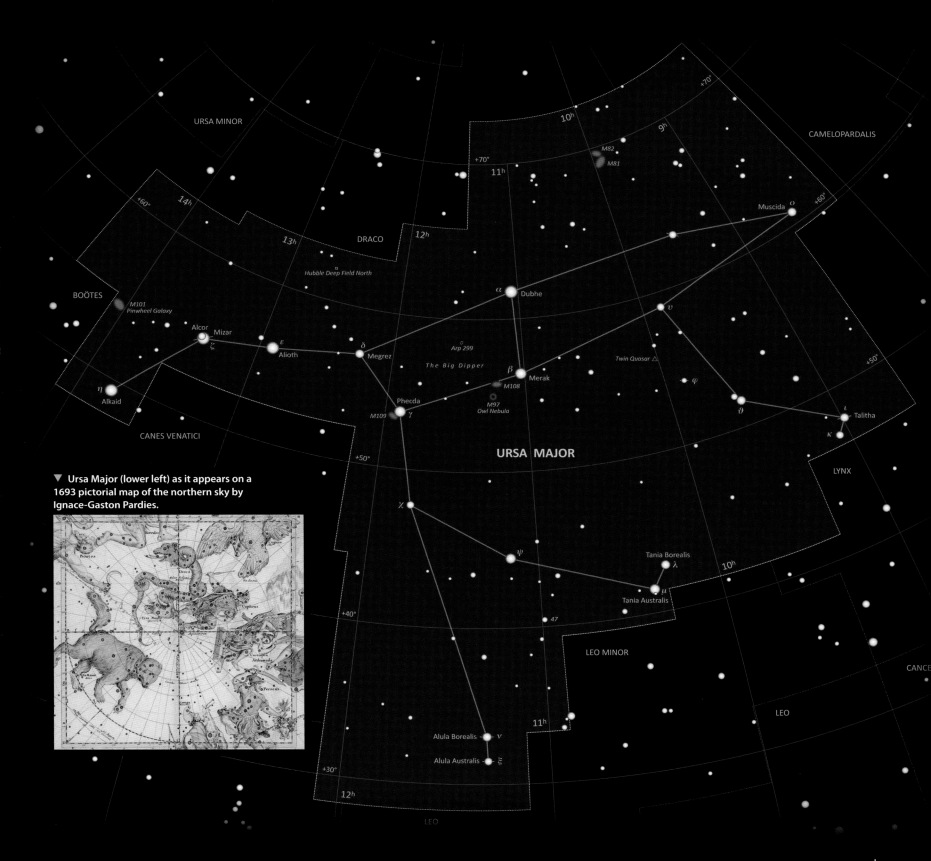

URSA MINOR

CAMELOPARDALIS

+70°

10ʰ

9ʰ

M82

M81

Muscida

ο

+60°

11ʰ

+70°

+60°

14ʰ

DRACO

12ʰ

Muscida

BOÖTES

13ʰ

Hubble Deep Field North

α Dubhe

υ

M101
Pinwheel Galaxy

Alcor

Mizar

ξ

ε

Alioth

δ

Megrez

Arp 299

The Big Dipper

β

Merak

M108

Twin Quasar △

φ

+50°

η

Alkaid

Phecda

M97
Owl Nebula

ϑ

ι

Talitha

M109

γ

κ

CANES VENATICI

URSA MAJOR

LYNX

+50°

▼ **Ursa Major (lower left) as it appears on a
1693 pictorial map of the northern sky by
Ignace-Gaston Pardies.**

χ

ψ

Tania Borealis

λ

10ʰ

μ

Tania Australis

47

+40°

LEO MINOR

CANCE

11ʰ

LEO

Alula Borealis — ν

Alula Australis — ξ

+30°

12ʰ

LEO

◄ The Pinwheel galaxy is seen almost exactly face-on. This image was obtained by the Hubble Space Telescope.

▼ The original Hubble Deep Field image shows almost 3,000 extremely remote galaxies in a small field in Ursa Major.

▶ At a distance of some 13.4 billion light-years in Ursa Major, GN-z11 (magnified) is the most distant galaxy known to date.

▼ 47 Ursae Majoris b (foreground) was the first exoplanet found in a wide orbit around a sun-like star.

1774   German astronomer Johann Bode discovers two galaxies in Ursa Major: M81 (also known as Bode's galaxy) and neighboring M82 (nicknamed the Cigar galaxy). M81 is a grand-design spiral galaxy at a distance of approximately 12 million light-years and visible with binoculars. M82 is a so-called starburst galaxy with an explosive core.

1781   In Paris, Pierre Méchain discovers the Pinwheel galaxy (M101)—a beautiful spiral seen almost exactly face-on. It is at least 50 percent larger than our Milky Way; its distance is about 21 million light-years.

1857   At Harvard College Observatory, John Whipple and George Bond succeed in obtaining the first photograph of a double star, Mizar and Alcor. The famous pair, at 83 light-years from Earth, is actually a system of six stars: Mizar consists of two close binaries orbiting each other; Alcor itself is also a binary star.

1979   Astronomers at the Kitt Peak National Observatory in Arizona discover two distant quasars in Ursa Major (QSO 0957+561A/B) that are very similar and close together in the sky. In fact, there is only one remote quasar, but its light is split into two separate images by the gravity of a fainter foreground galaxy. The Twin Quasar is the first convincing example of gravitational lensing—an effect predicted by Albert Einstein's theory of general relativity.

1995   For ten consecutive days in December, the Hubble Space Telescope is aimed at a small, apparently empty region of sky in Ursa Major, in order to take a representative census of extremely remote galaxies. The resulting Hubble Deep Field image reveals almost three thousand galaxies, including some of the most distant objects ever observed.

◄ Combining optical, X-ray, and infrared observations of M82 reveals the explosive nature of this galaxy.

1996   The first long-period extrasolar planet is discovered by American astronomers Geoff Marcy and Paul Butler, around the sun-like star 47 Ursae Majoris. 47 UMa b is a massive giant planet orbiting its parent star once every 2.95 years. In 2002 and 2010, two more planets are found at even larger distances from the star.

2005   Astronomers witness the first stage of an outburst in the peculiar galaxy Arp 299 in Ursa Major, at a distance of almost 150 million light-years from Earth. Subsequent observations at radio wavelengths reveal that this is a so-called tidal disruption event: A star is being ripped apart by the tidal forces of a supermassive black hole.

2014   While training students at a small observatory in northern London, British astronomer Steve Fossey serendipitously discovers a supernova in the Cigar galaxy (M82). At a distance of just 12 million light-years, SN 2014J is one of the closest supernovae seen for decades.

2016   Follow-up observations of an extremely faint object in the Hubble Deep Field North reveal it to be a primordial galaxy at a distance of 13.4 billion light-years. As of 2018, GN-z11 is the most distant galaxy known.

► The Twin Quasar— the result of gravitational lensing—is in the center of this Hubble image. The lensing galaxy can be seen around one of the images.

URSA MINOR

PASSPORT

Latin name: Ursa Minor	**Area:** 255.9 square degrees
English name: Little Bear	**Number of naked-eye stars:** 39
Genitive: Ursae Minoris	**Bordering constellations:** Cepheus, Camelopardalis, Draco
Abbreviation: UMi	**Best visibility:** April–May, north of 5° north
Origin: Ptolemy	

URSA MINOR (THE LITTLE BEAR) is the most northerly constellation in the sky. It contains the north celestial pole, which lies very close to the bright star Polaris (Alpha Ursae Minoris, or α UMi). Polaris, also known as the Pole Star, is an important navigational aid in the northern hemisphere: It always marks the direction of true north, while its altitude above the horizon equals the geographical latitude of the observer.

Like Ursa Major (the Great Bear), Ursa Minor has a characteristic dipper shape, although the handle of the dipper curves away in the opposite direction. Also, its seven stars are fainter than the stars of the Big Dipper.

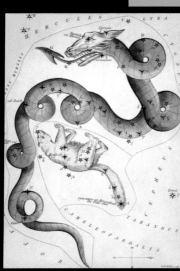

▲ Ursa Minor and Draco, as they appear in *Urania's Mirror*, an 1825 set of constellation cards.

◀ **Illustration of Polaris and its two companion stars, Polaris Ab and Polaris B. The multiple star system is some 430 light-years away.**

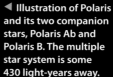

◀ **On a long-exposure photo, like this one by Frederick Steiling, it is evident that Polaris (center) is very close to the north celestial pole.**

▼ NGC 6217 is one of the few
relatively bright galaxies that
can be found in Ursa Minor.

▶ The Pole Star (upper left)
can be found by extending
the right side of the bowl of
the Big Dipper (lower right).

TIMELINE

1797   From his home in Slough, England, William Herschel discovers the barred spiral galaxy NGC 6217, at a distance of approximately 67 million light-years in Ursa Minor.

1911   Danish astronomer Ejnar Hertzsprung confirms the suspected variability of Polaris. The giant star (about forty times as large and some 1,400 times as luminous as the sun) is a Cepheid variable, pulsating once every four days or so.

1955   Albert Wilson of the Lowell Observatory in Arizona studies photographic plates from the Palomar Observatory Sky Survey and discovers the Ursa Minor dwarf galaxy, a loose collection of extremely faint stars, some 225,000 light-years away.

1980   In *The Restaurant at the End of the Universe* (the second book in the Hitchhiker's Guide to the Galaxy series), Douglas Adams describes the (fictional) planet Ursa Minor Beta. Its surface solely consists of tropical coastline, and it is always Saturday afternoon.

2004   Astronomers report evidence that Polaris was 2.5 times fainter than it is now when observed by Ptolemy in the second century. This would constitute an unprecedented fast evolutionary change.

VELA

PASSPORT

Latin name: Vela	**Area:** 499.6 square degrees
English name: Sails	**Number of naked-eye stars:** 214
Genitive: Velorum	**Bordering constellations:** Pyxis, Puppis, Carina, Centaurus, Antlia
Abbreviation: Vel	
Origin: de Lacaille	**Best visibility:** January–March, south of 30° north

▶ **NGC 2547 is a group of young, sparkling stars in Vela.**

▼ **WASP-19b has clouds of titanium oxide in its thick atmosphere. A year on this planet only lasts 18.9 hours.**

VELA (THE SAILS) is a constellation in the southern sky that used to be part of Argo Navis (Ship *Argo*), before French astronomer Nicolas Louis de Lacaille divided this extremely large constellation into Carina (the Keel), Puppis (the Stern), and Vela. It lies in the band of the Milky Way, and contains quite a lot of nebulae and star clusters.

Vela was also home to a bright supernova explosion that must have occurred some twelve thousand years ago. The expanding remains of this stellar blast are still visible, as is the rapidly spinning Vela pulsar—the collapsed core of the exploding star.

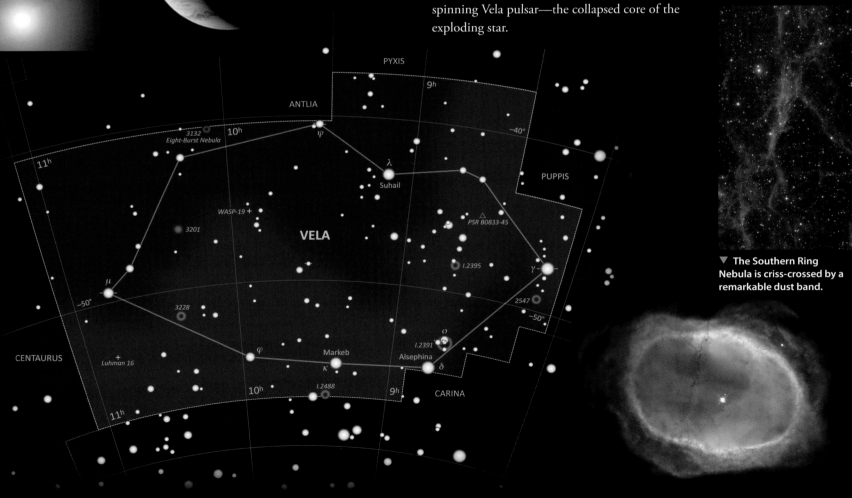

▼ **The Southern Ring Nebula is criss-crossed by a remarkable dust band.**

▼ The sun, at just 6.6 light-years away, appears as a bright star (upper left) in this artist's impression of Luhman 16, the binary brown dwarf in Vela.

1751   Nicolas Louis de Lacaille, who also introduced the constellation, discovers a beautiful, compact open star cluster, NGC 2547. Its distance is approximately 1,500 light-years.

1835   From his observatory in Cape Town, John Herschel discovers the planetary nebula NGC 3132. Because of its shape, it is nicknamed the Eight-Burst Nebula, or the Southern Ring Nebula. The slowly expanding shell of gas is some 2,000 light-years away, at the border of Vela and Antlia (the Air Pump).

1968   Australian astronomers study the Vela pulsar (PSR B0833-45), which spins around its axis every 11.2 seconds, and suggest it is related to the Vela Supernova Remnant. This is the first evidence for a relation between supernovae and pulsars.

2009   The Wide Angle Search for Planets program (WASP) discovers an exoplanet with an orbital period of just 18.9 hours—the shortest known at the time. It orbits the star WASP-19 in Vela. The Jupiter-sized planet has a thick atmosphere containing water vapor and titanium oxide.

2013   American astronomer Kevin Luhman discovers a binary brown dwarf (Luhman 16) in Vela at a distance of just 6.6 light-years—the closest brown dwarfs known.

▶ X-ray image of the Vela Pulsar, which spews out a jet of energetic particles into space.

◀ The Vela Supernova Remnant is all that's left of a titanic stellar explosion that occurred some 12,000 years ago.

VIRGO

Latin name: Virgo	**Area:** 1,294.4 square degrees
English name: Virgin	**Number of naked-eye stars:** 169
Genitive: Virginis	**Bordering constellations:** Coma Berenices, Leo, Crater, Corvus,
Abbreviation: Vir	Hydra, Libra, Serpens Caput, Boötes
Origin: Ptolemy	**Best visibility:** March–May, between 65° north and 75° south

VIRGO (THE VIRGIN) is the second-largest constellation in the night sky, after Hydra (the Sea Serpent). Its brightest star, Spica (Alpha Virginis, or α Vir) is one of the brightest stars in the northern hemisphere. The constellation is easy to find, by extending the curved handle of the Big Dipper, first to Arcturus (Alpha Boötis, or α Boo) and then on to Spica.

Virgo is associated with Dike, the Greek goddess of justice, but also with Demeter, the goddess of harvest and fertility. In ancient representations, Virgo is often depicted with an ear of grain in her hand—actually, the name *Spica* means "ear of grain."

The constellation is home to the relatively close Virgo Cluster and contains a large number of galaxies. Since Virgo is also part of the zodiac (the twelve constellations in which the sun, moon, and planets can be found), many solar system discoveries and spaceflight events occurred within the constellation's boundaries.

◄ Pluto's large moon Charon, discovered in 1978, was visited by NASA's New Horizons spacecraft just 37 years later, in 2015.

▼ On July 20, 1969, astronauts left their first footprints on the moon, which was in Virgo that day.

▲ The planets of pulsar PSR B1257+12 are battered by energetic particles and radiation of its rapidly spinning and highly magnetized parent star (upper left).

▲ At infrared wavelengths, the slightly warped dust band of the Sombrero galaxy (M104) becomes very prominent.

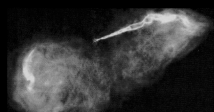

◄ Radio telescopes reveal fine detail in the energetic jet of the galaxy M87.

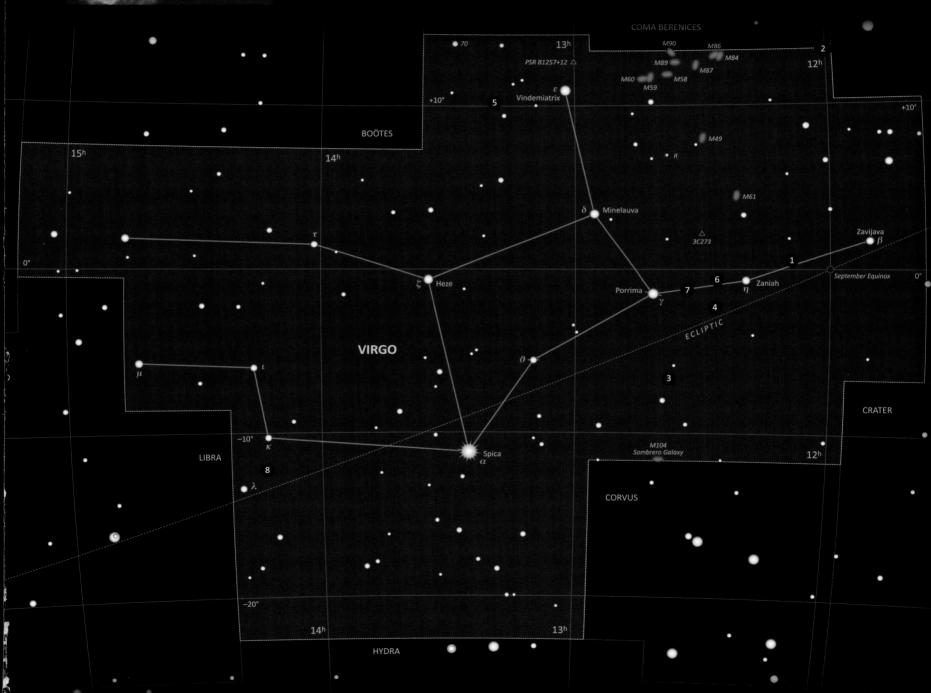

COMA BERENICES

BOÖTES

VIRGO

LIBRA

HYDRA

CORVUS

CRATER

70

13ʰ

PSR B1257+12 △

+10°

14ʰ

15ʰ

ε
Vindemiatrix

5

δ Minelauva

τ

ζ
Heze

μ

ι

κ

λ

8

0°

−10°

−20°

M90   M86
                     M84
M89   M87
M60   M58
    M59

M49

R

M61

3C273 △

September Equinox

ECLIPTIC

Zavijava
β

1

6
η Zaniah

7
Porrima
γ

4

ϑ

3

Spica
α

M104
Sombrero Galaxy

12ʰ

12ʰ

+10°

0°

12ʰ

14ʰ   *13ʰ*

▼ The south pole of asteroid Vesta, as imaged by NASA's Dawn spacecraft.

▼ Viking 2 photo of the rocky desert of Utopia Planitia on Mars.

▶ Voyager 2 captured fine detail in the turbulent atmosphere of Saturn.

▲ Comet Shoemaker-Levy 9 left dark stains in Jupiter's atmosphere after impacting with the giant planet.

▶ Christiaan Huygens was the first to realize that Saturnus is surrounded by a thin, flat ring.

TIMELINE

1612 ❶ Galileo Galilei studies the planet Jupiter and its four major moons—in Virgo at the time—and notes a nearby star which later turns out to be the planet Neptune. Neptune is not discovered until 1846.

1767 French astronomer Pierre Méchain discovers M104, on the border of Virgo and Corvus (the Raven). Also known as the Sombrero galaxy, it has an extended halo of stars and globular clusters and a prominent dust band. The galaxy is some 30 million light-years away.

1807 ❷ On March 29, German astronomer Heinrich Olbers finds asteroid Vesta, in the extreme northwest corner of Virgo. It is only the fourth asteroid known, after Ceres, Pallas, and Juno. Back then, these solar system bodies were regarded as bona fide planets.

1918 American astronomer Heber Curtis discovers a strange linear jet, pointing away from the core of M87. At that time, the extragalactic nature of this "nebula" isn't even understood. We now know that the jet is produced by the supermassive black hole in this huge elliptical galaxy, which sits at the core of the Virgo Cluster, some 55 million light-years away.

1963 The first quasar (quasi-stellar object) is discovered in Virgo by Dutch-American astronomer Maarten Schmidt; 3C 273 is approximately 2.4 billion light-years away but is bright enough to be seen with amateur telescopes.

1969 ❸ The moon is in Virgo when Apollo 11 commander Neil Armstrong is the first person to set foot on another world, on July 20.

1976 ❹ The dishes of NASA's Deep Space Network are aimed at Virgo to pick up the first signals and photos from the Viking 2 lander as it touches down on Mars on September 3.

1978 ❺ At the United States Naval Observatory, James Christy discovers Charon, the largest moon of Pluto. At the time, Pluto is in the northern part of Virgo.

1980 ❻❼ The ringed planet Saturn is in Virgo at the time of the Voyager 1 flyby, on November 12. Eight months later, when Voyager 2 arrives, it is still within the constellation's boundaries.

1992 American radio astronomers discover the first extrasolar planets, orbiting the pulsar PSR B1257+12, at a distance of some 2,300 light-years in Virgo. A third planet is found in 1994. The three planets are officially named Draugr, Poltergeist, and Phobetor.

1994 ❽ Telescopes on the ground and in space are aimed at Virgo to witness the predicted impacts of the fragments of comet Shoemaker-Levy 9 in the atmosphere of Jupiter.

1996 A Jupiter-like planet is discovered orbiting the star 70 Virginis. It is one of the first extrasolar planets found around a sun-like star; 70 Vir b has an orbital period of 117 days.

VOLANS

PASSPORT

Latin name: Volans	**Area:** 141.4 square degrees
English name: Flying Fish	**Number of naked-eye stars:** 31
Genitive: Volantis	**Bordering constellations:** Carina, Pictor, Dorado, Mensa, Chamaeleon
Abbreviation: Vol	
Origin: Keyser and de Houtman	**Best visibility:** December–February, south of 10° north

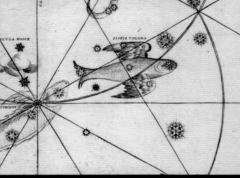

▼ **Piscis Volans, as it was originally called, appears in Johann Bayer's 1603 star atlas *Uranometria*.**

▲ **At a distance of some 200 million light-years in Volans, galaxy ESO 60-24 is seen almost exactly edge-on.**

VOLANS (THE FLYING FISH) is one of the twelve new constellations that were suggested by Dutch sailors Pieter Dirkszoon Keyser and Frederick de Houtman, who named them after the wondrous creatures they encountered during their journeys. The constellation is not easy to recognize; it lies between the Large Magellanic Cloud and the bright star Miaplacidus (Beta Carinae, or β Car).

Volans occupies a rather poor region in the southern sky, and contains just a handful of interesting objects, the most famous of which is the Meathook galaxy (NGC 2442).

▶ **Bright blue stars mark the spiral arms of NGC 2397.**

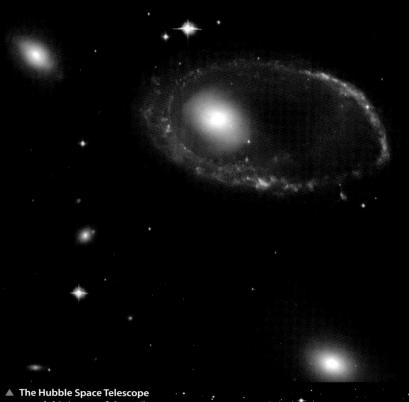

▲ **The Hubble Space Telescope captured this image of the striking ring galaxy AM 0644-741.**

▶ **The Meathook galaxy (NGC 2442) is some 50 million light-years away.**

TIMELINE

1598   The new constellation is formally introduced by Flemish astronomer Petrus Plancius on a small celestial globe made by Jodocus Hondius in Amsterdam.

1834   While observing the southern sky from Cape Town, English astronomer John Herschel (son of William) discovers NGC 2442. The Meathook galaxy, as it is generally known, has an asymmetric shape, most likely due to a gravitational encounter with a smaller galaxy in the distant past.

1835   John Herschel also discovers NGC 2397, a classical spiral galaxy at a distance of approximately 60 million light-years in Volans.

1960   American astronomers Eric Lindsay and Harlow Shapley discover a remarkable ring-shaped galaxy in Volans, at a distance of 300 million light-years. Like the Cartwheel galaxy in Sculptor, AM 0644-741 is the result of a head-on collision of two galaxies. The ring is 150,000 light-years wide and contains many young stars.

2002   A Jupiter-sized planet is discovered in a tight orbit around the old, sun-like star HD 76700 in Volans.

VULPECULA

PASSPORT

Latin name: Vulpecula		**Area:** 268.2 square degrees	
English name: Little Fox		**Number of naked-eye stars:** 68	
Genitive: Vulpeculae		**Bordering constellations:** Cygnus, Lyra, Hercules, Sagitta, Delphinus, Pegasus	
Abbreviation: Vul			
Origin: Hevelius		**Best visibility:** July–August, north of 60° south	

VULPECULA (THE LITTLE FOX) was introduced as Vulpecula et Anser (the Little Fox with the Goose) by Polish astronomer Johannes Hevelius in his 1687 star atlas *Firmamentum Sobiescianum.* It is a relatively small and faint constellation between the bright stars Deneb (Alpha Cygni, or α Cyg) and Altair (Alpha Aquilae, or α Aql).

Vulpecula lies in the Milky Way and contains quite a number of star clusters and nebulae. By far the most famous is the Dumbbell Nebula (M27). Vulpecula is also home to the first pulsar and the first millisecond pulsar ever discovered.

◄ **Exoplanet HD 189733b is so close to its hot parent star that its atmosphere is slowly eroding away.**

▲ **Small knots of gas can be seen in this close-up of the Dumbbell Nebula, obtained by the Hubble Space Telescope.**

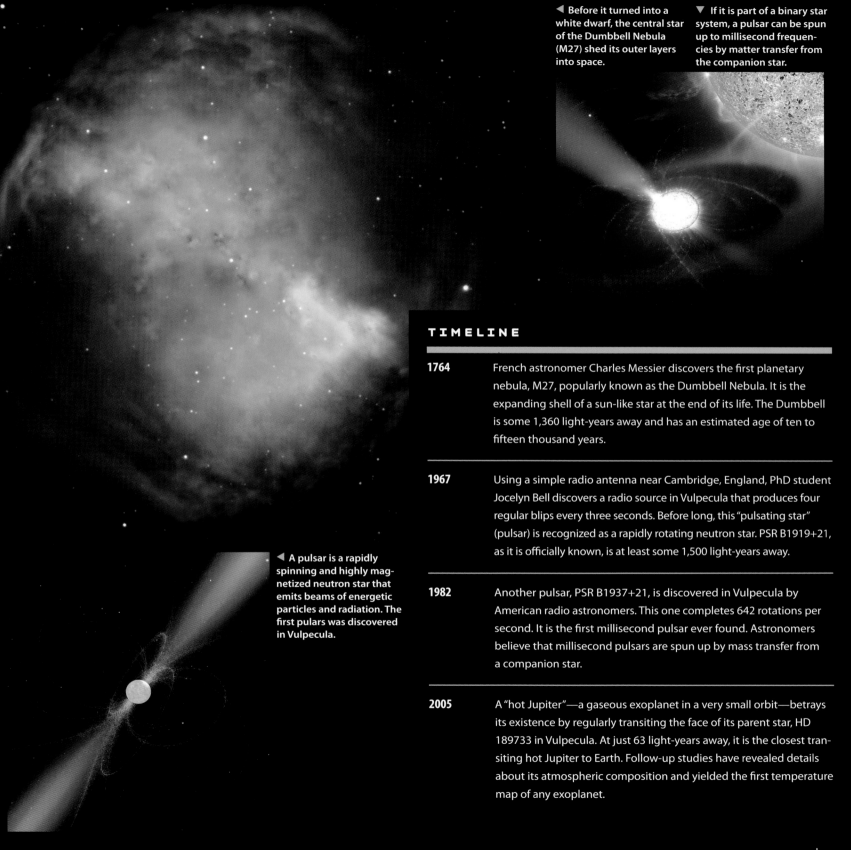

◄ Before it turned into a white dwarf, the central star of the Dumbbell Nebula (M27) shed its outer layers into space.

▼ If it is part of a binary star system, a pulsar can be spun up to millisecond frequencies by matter transfer from the companion star.

◄ A pulsar is a rapidly spinning and highly magnetized neutron star that emits beams of energetic particles and radiation. The first pulars was discovered in Vulpecula.

TIMELINE

1764   French astronomer Charles Messier discovers the first planetary nebula, M27, popularly known as the Dumbbell Nebula. It is the expanding shell of a sun-like star at the end of its life. The Dumbbell is some 1,360 light-years away and has an estimated age of ten to fifteen thousand years.

1967   Using a simple radio antenna near Cambridge, England, PhD student Jocelyn Bell discovers a radio source in Vulpecula that produces four regular blips every three seconds. Before long, this "pulsating star" (pulsar) is recognized as a rapidly rotating neutron star. PSR B1919+21, as it is officially known, is at least some 1,500 light-years away.

1982   Another pulsar, PSR B1937+21, is discovered in Vulpecula by American radio astronomers. This one completes 642 rotations per second. It is the first millisecond pulsar ever found. Astronomers believe that millisecond pulsars are spun up by mass transfer from a companion star.

2005   A "hot Jupiter"—a gaseous exoplanet in a very small orbit—betrays its existence by regularly transiting the face of its parent star, HD 189733 in Vulpecula. At just 63 light-years away, it is the closest transiting hot Jupiter to Earth. Follow-up studies have revealed details about its atmospheric composition and yielded the first temperature map of any exoplanet.

IMAGE CREDITS

Pg. 1, Wil Tirion; 2, Wil Tirion; 10–11, Collection Henk Bril; 12, Collection Henk Bril; 14, Rob van Gent (Utrecht University); 15 (UR), Wikimedia Commons; 15 (LR), Wikimedia Commons; 16–17, John Colosimo/ESO; 16 (UC), Wikimedia Commons; 16 (ML), Wikimedia Commons; 16 (LC), Wikimedia Commons; 17 (LR), Wikimedia Commons; 18, Rob van Gent (Utrecht University); 19, Utrecht University; 20–21, Collection Henk Bril; 20 (LC), Wikimedia Commons; 22, ESO/L. Calçada; 23, NASA; 24, ESO/L. Calçada; 25 (UC), NASA/ESA/C.R. O'Dell (Vanderbilt University)/M. Meixner, P. McCullough, G. Bacon (STScI); 25 (LL), NASA/ESA/Erich Karkoschka (University of Arizona); 26 (MR), NASA/ESA; 26 (LR), NASA/SDO; 27, ESO/M. Kornmesser; 28 (LL), Collection Henk Bril; 28 (LR), ESO/Y. Beletsky; 29 (UR), Collection Henk Bril; 29 (MR), ESO/Y. Beletsky; 30–31, Collection Henk Bril; 33, Wil Tirion; 35, Wil Tirion; 37, Wil Tirion; 38–39, Collection Henk Bril; 40 (LL), Bodleian Library, Oxford; 41 (UR), NASA/JPL-Caltech/P. Barmby (Harvard-Smithsonian CfA); 41 (MR), NASA/ESA/Thomas M. Brown, Charles W. Bowers, Randy A. Kimble, Allen V. Sweigart (Goddard Space Flight Center)/Henry C. Ferguson (STScI); 41 (LL), MarioProtIV; 41 (LR), NASA/ESA/Hubble Heritage Team (STScI/AURA); 42 (UL), Robert Gendler; 42 (UR), NASA/ESA/Andrew Hamilton (University of Colorado)/Robert Fesen (Dartmouth College); 42 (MR), NASA/Walt Feimer; 42 (LR), NASA/ESA/STScI/D. Geisler (Universidad de Concepción); 43, NASA/ESA/Z. Levay, R. van der Marel (STScI)/A. Mellinger; 44 (UR), ALMA (ESO/NAOJ/NRAO)/F. Kerschbaum; 44 (ML), NASA/ESA/Nick Rose; 44 (LR), NASA/ESA; 45 (LL), NASA/ESA; 45 (MC), Wikimedia Commons; 45 (UR), NASA/ESA; 46 (MR), NASA/ESA/E. Karkoschka; 46 (LR), Wikimedia Commons; 47 (UC), NASA/G. Bacon; 47 (MR), ESO; 47 (LC), NRAO/AUI/NSF/D. Berry/ALMA; 48 (LL), Harvard-Smithsonian CfA; 48 (LR), Wikimedia Commons; 49 (UL), NASA/JPL-Caltech/University of Arizona; 49 (ML), NASA/CXC/ESA/NAOJ; 49 (UR), NASA/JPL-Caltech; 50 (UR), NASA/Goddard Space Flight Center; 50 (LR), B. Saxton, NRAO/AUI/NSF from data provided by F. Lockman; 51 (UL), Ukrainian National Observatory; 51 (ML), Zina Deretsky/National Science Foundation; 51 (LL), ESO/L. Calçada; 52 (LL), ESO; 52 (LC), Matt Bobrowsky (Orbital Sciences Corporation)/NASA/ESA; 52 (LR), ESA/NASA & Hubble; 53 (UL), Göran Nilsson/The Liverpool Telescope; 53 (UC), Wikimedia Commons; 53 (LR), Celestia Team; 54 (MC), NASA/JPL-Caltech; 54 (LL), CFHT; 55 (UL), NASA/ESA/HST Frontier Fields team (STScI); 55 (UR), Linda Hall Library/Wikimedia Commons; 55 (LL), Jingchuan Yu/Beijing Planetarium; 56 (UC), ESO; 56 (LL), ESA/NASA & Hubble; 57 (UR), Volker Springel/Millennium Simulation Project; 57 (MR), ESO/L. Calçada; 57 (LR), ESO/M. Kornmesser; 58 (UL), ESO Online Digitized Sky Survey; 58 (LL), ESO; 58 (LR), Wikimedia Commons; 59 (UL), Wikimedia Commons; 59 (UC), NASA/ESA/M. Postman, D. Coe (STScI)/CLASH team; 59 (LR), NASA/ESA/A. V. Filippenko (University of California, Berkeley)/P. Challis (Harvard-Smithsonian CfA) et al.; 60 (UR), NASA/JPL; 60 (MR), NASA/JPL; 60 (LR), NASA/JPL; 61 (UL), Bob Franke/Focal Pointe Observatory; 61 (MR), NASA/JPL-Caltech; 62, X-ray: NASA/CXC/Caltech/P. Ogle et al./Optical: NASA/STScI/R. Gendler/Infrared: NASA/JPL-Caltech/Radio: NSF/NRAO/VLA; 63 (UL), Wikimedia Commons; 63 (ML), NASA/ESA/S. Beckwith (STScI)/Hubble Heritage Team (STScI/AURA); 63 (LL), NASA/ESA/STScI/A. Sarajedini (University of Florida); 63 (LR), St. Andrews Special Collections; 64 (LL), 2MASS/UMass/IPAC-Caltech/NASA/NSF; 64 (LC), NASA/ESA/H. Bond (STScI)/M. Barstow (University of Leicester); 65 (UL), Nicolas Martin, Rodrigo Ibata (Observatoire de Strasbourg); 65 (LL), ESA/Hubble & NASA/Luca Limatola; 65 (LR), NASA/ESA/G. Bacon (STScI); 66 (UC), Danielle Futselaar/METI; 66 (LL), USNO; 66 (LR), Caltech; 67 (ML), NASA/ESA; 67 (LL), Wikimedia Commons; 67 (LC), NASA/JPL/Space Science Institute; 68 (LC), NASA/ESA/M. Livio, Hubble 20th Anniversary Team (STScI); 68 (LR), ESO; 69 (UL), NASA/ESA/Hubble Heritage Team (STScI/AURA)/A. Nota (ESA/STScI)/Westerlund 2 Science Team; 69 (LR), X-ray: NASA/CXC/Harvard-Smithsonian CfA/M. Markevitch et al./Optical: NASA/STScI/Magellan/University of Arizona/D. Clowe et al./Lensing Map: NASA/STScI/ESO WFI; 70 (LL), University of Toronto; 70 (LR), NASA/JPL-Caltech/STScI/CXC/SAO; 71 (UR), ESA/Hubble & NASA/S. Smartt (Queen's University Belfast); 71 (LL), G. S. Hughes, S. Maddox (RGO) et al., Isaac Newton Telescope; 71 (LR), NASA/JPL-Caltech; 73 (UL), Erik Lernestal/Skokloster Castle/Wikimedia Commons; 73 (UC), ESA/Hubble & NASA; 73 (LC), ESO/M. Kornmesser; 74 (UL), Romano Corradi (Instituto de Astrofisica de Canarias)/Mario Livio (STScI)/Ulisse Munari (Osservatorio Astronomico di Padova-Asiago)/Hugo Schwarz (Nordic Optical Telescope)/NASA/ESA; 74 (UR), ESO; 74 (LC), ESO/INAF-VST/OmegaCAM/A. Grado, L. Limatola (INAF-Capodimonte Observatory); 75 (LL), NASA/ESA/Hubble SM4 ERO Team; 75 (LR), ESA/Hubble & NASA; 76 (UL), Geert Barentsen, Jorick Vink (Armagh Observatory)/IPHAS Collaboration; 76 (LL), ESO; 76 (LR), NASA/JPL-Caltech; 77 (UR), Keith Seabridge/Openplaques.org; 77 (LL), ESA/NASA/JPL-Caltech/Whitman College; 79 (UR), ESO; 79 (LL), NASA/ESA/P. van Dokkum (Yale University); 79 (LR), David Jewitt/Jane Luu; 80 (UL), NASA/JPL-Caltech/C. Martin (Caltech)/M. Seibert (OCIW); 80 (LL), NASA/JPL-Caltech/R. Hurt (SSC); 80 (LR), ESO/L. Calçada/Nick Risinger (skysurvey.org); 81 (UR), NASA/ESA/Kevin Luhman (Pennsylvania State University)/Judy Schmidt; 81 (LL), ESO; 81 (LR), NASA/ESA/Howard Bond (STScI); 82 (UL), NASA/ESA/Judy Schmidt; 82 (UC), ASA/JPL-Caltech/UCLA; 82 (LL), X-ray: NASA/CXC/University of Wisconsin-Madison/S. Heinz et al./Optical: DSS; 83 (MC), Barry Lawrence Ruderman Antique Maps Inc.; 83 (MR), Digital Sky Survey/WikiSky.org; 83 (LR), NASA/ESA/Fabian RRRR/Wikimedia Commons; 84 (UL), NASA/ESA/Holland Ford (JHU)/ACS Science Team; 84 (LR), Rogelio Bernal Andreo; 85 (UL), NASA/ESA/Hubble; 85 (UR), ESO/A. Roquette; 85 (LL), NASA/ESA/Digitized Sky Survey 2/Davide De Martin; 86 (MR), Loke Kun Tan (StarryScapes.com); 86 (LC), University of Oklahoma/History of Science Collections; 86 (LR), NASA/ESA/Hubble; 87 (ML), David Ritter/Wikimedia Commons; 87 (LL), Aldaron/Wikimedia Commons; 87 (LR), Stefan Heutz/Wolfgang Ries/Michael Breite; 88 (UL), Robert Gendler; 88 (MR), ESA/Hubble & NASA; 88 (LC), Harvard Map Collection/Wikimedia Commons; 89 (MR), NASA/CXC/University of Michigan/R. C. Reis et al./Optical: NASA/STScI; 89 (LL), Carnegie-Irvine Galaxy Survey; 90 (UR), ESO/Y. Beletsky; 90 (LL), Wikimedia Commons; 91 (UL), NASA/ESA/Jesús Maíz Apellániz (Instituto de Astrofísica de Andalucía); 91 (MC), ESO/Digitized Sky Survey 2/Davide De Martin; 91 (LL), ESO; 92 (UR), T. A. Rector (University of Alaska Anchorage)/WIYN/NOAO/AURA/NSF; 92 (LL), Miodrag Sekulic/Wikimedia Commons; 92 (LR), NASA/JPL-Caltech; 93 (UL), NASA/Tim Pyle; 94 (UL), NRAO/AUI; 94 (UR), NASA/

JPL-Caltech/Harvard-Smithsonian CfA; 94 (LL), Martin Kornmesser/ESA/ECF; 95 (UC), Wikimedia Commons; 95 (UR), NASA/ESA/Judy Schmidt; 95 (LR), ESA/Hubble & NASA; 96 (UR), ESA/NASA/Martino Romaniello (ESO); 96 (LC), ESO; 97 (UC), NASA/ESA/F. Paresce (INAF-IASF)/R. O'Connell (University of Virginia, Charlottesville)/Wide Field Camera 3 Science Oversight Committee; 97 (ML), ESO; 97 (LL), ESA/Hubble & NASA; 98 (UR), Gregg Ruppel; 98 (LL), NASA/ESA/HEIC/Hubble Heritage Team (STScI/AURA); 98 (LR), NASA/ESA/ Holland Ford (JHU)/ACS Science Team; 99 (UL), USNO; 100 (UL), Nordic Optical Telescope/ Romano Corradi (Isaac Newton Group); 100 (LL), NASA/ESA/Hubble Heritage Team (STScI/ AURA); 100 (LR), NASA/Kepler Mission/Dana Berry; 101 (UL), Harvard Map Collection/ Wikimedia Commons; 101 (LL), NASA/JPL-Caltech; 101 (LR), Sloan Digital Sky Survey; 102– 103, NASA/STScI Digitized Sky Survey/Noel Carboni; 103 (MR), NASA/JPL-Caltech; 103 (LR), NASA/ESA Hubble Heritage Team (STScI/AURA); 104 (LL), Fred the Oyster/Wikimedia Commons; 104 (LR), NASA/ESA/D. Harvey (École Polytechnique Fédérale de Lausanne)/R. Massey (Durham University)/HST Frontier Fields team; 105 (UL), NASA/WMAP Collaboration; 105 (LL), ESO/IDA/Danish 1.5 m/R. Gendler/A. Hornstrup; 106 (UL), NASA/ESA/N. Pirzkal (ESA/STScI)/HUDF Team (STScI); 106 (UR), ESO/Aniello Grado/Luca Limatola; 106 (LL), ESO/ Digitized Sky Survey 2; 106 (LR), ESO/R. Gendler; 107 (UL), NRAO/AUI/J. M. Uson; 107 (LR), ESO/IDA/Danish 1.5 m/R. Gendler/J.-E. Ovaldsen/C. Thöne/C. Feron; 108 (UR), NASA/ESA/ Andrew Fruchter (STScI)/ERO team (STScI/ST-ECF); 108 (LR), ESA/NASA/JPL/University of Arizona; 109 (MR), ESO; 109 (LL), Lowell Observatory; 110 (LL), NASA/JPL/University of Arizona/University of Idaho; 110 (LC), WikiSky; 110 (LR), NASA/CXC/GWU/N. Klingler et al.; 111 (UR), ESO; 111 (LL), Wikimedia Commons; 111 (LR), NASA/ESA/J. Lotz (STScI); 112 (UL), NASA/ESA/S. Baum, C. O'Dea (RIT)/R. Perley, W. Cotton (NRAO/AUI/NSF)/Hubble Heritage Team (STScI/AURA); 112 (MC), ESA/Hubble & NASA; 113 (UR), USNO; 113 (LR), Wikimedia Commons; 114 (UR), NASA/STScI/WikiSky; 114 (LC), ESO/INAF-VST/OmegaCAM/OmegaCen/ Astro-WISE/Kapteyn Institute; 115 (UC), Wikimedia Commons/Hubble Space Telescope Legacy Archive; 115 (MR), ESA/Hubble & NASA; 115 (LR), ESA/Hubble & NASA/D. Calzetti (UMass)/LEGUS Team; 116 (UC), NASA/ESA/Hubble Heritage Team (STScI/AURA)/W. Blair (STScI/JHU), Carnegie Institution of Washington (Las Campanas Observatory)/NOAO; 117 (UL), S. Andrews (Harvard-Smithsonian CfA)/B. Saxton (NRAO/AUI/NSF)/ALMA (ESO/NAOJ/ NRAO); 118 (MC), ESA/Hubble & NASA; 118 (MR), M. Blanton/SDSS; 118 (LL), NASA/ESA; 119 (UL), Robin Dienel/Carnegie Institution for Science; 119 (UR), 2MASS/UMass/IPAC-Caltech/ NASA/NSF; 120 (UR), ESO/L. Calçada; 120 (LL), Wikimedia Commons/Hubble Space Telescope Legacy Archive; 120 (LR), ESA/Hubble & NASA/Judy Schmidt; 121 (UL), NASA/ ESA/W. Harris (McMaster University, Ontario); 121 (UC), ESO; 121 (UR), Gemini Observatory/ Jon Lomberg; 122, Wil Tirion; 123 (UR), W. Steffen (UNAM)/J. L. Gómez (IAA); 123 (MR), ESA/ Hubble & NASA; 123 (LR), Casey Reed/NASA; 125 (UC), NASA/ESA/S. Rodney (John Hopkins University)/FrontierSN team/T. Treu (University of California Los Angeles)/P. Kelly (University of California Berkeley)/GLASS team/J. Lotz (STScI)/Frontier Fields team/M. Postman (STScI)/ CLASH team/Z. Levay (STScI); 125 (MC), NASA/ESA/Hubble Heritage (STScI/AURA)/ESA-Hubble Collaboration/Davide De Martin/Robert Gendler; 125 (LL), ESA/Hubble & NASA; 126 (LR), NASA/JPL; 127 (UC), NASA/JPL-Caltech; 127 (LL), NASA/JPL/SSI/Emily Lakdawalla; 127 (LR), NASA/JPL-Caltech/SwRI/ASI/INAF/JIRAM; 128 (UR), ESA/Hubble, NASA; 128 (MR), NASA/ESA/A. Riess (STScI); 128 (LR), NASA/ESA/William Keel (University of Alabama, Tuscaloosa)/Galaxy Zoo team; 129 (UR), WikiSky; 129 (MC), S. Kulkarni (Caltech)/D. Golimowski (JHU)/NASA/ESA; 129 (LL), NASA/ESA/S. Djorgovski (Caltech)/F. Ferraro (University of Bologna); 130 (UR), Wikimedia Commons; 131 (UR), NASA/JPL; 131 (MR), NASA/JPL-Caltech/UMD; 131 (LL), Wikimedia Commons/San Esteban; 131 (LR), ESO; 132 (LL), X-ray: NASA/CXC/Rutgers/G. Cassam-Chena, J. Hughes et al./Radio: NRAO/AUI/NSF/ GBT/VLA/Dyer, Maddalena & Cornwell; Optical: Middlebury College/F. Winkler/NOAO/ AURA/NSF/CTIO Schmidt/DSS; 132 (LC), NAOJ; 132 (LR), ESO/H. Avenhaus et al./DARTT-S collaboration; 133 (UL), Wikimedia Commons; 133 (UR), ESA/NASA/Robert Fosbury (ESA/ ST-ECF); 133 (LL), ESA/Hubble & NASA; 134 (UR), ESA/NASA/Gilles Chapdelaine; 134 (LL), NASA/ESA/Hubble Heritage Team (AURA/STScI); 134–135, ESO/M. Kornmesser; 135 (UR), NASA/JPL-Caltech; 135 (ML), NASA/Ames/JPL-Caltech; 136 (UC), NASA/CXC/SAO; 136 (UR), Eckhard Slawik (e.slawik@gmx.net); 136 (ML), NASA/ESA/Hubble Legacy Archive; 137 (UR), NASA/ESA/G. Bacon (STScI); 137 (MR), NASA/ESA/J. R. Graham, P. Kalas (University of California, Berkeley)/B. Matthews (Hertzberg Institute of Astrophysics); 137 (LL), Wikimedia Commons/Harvard Map Collection; 138, ESO; 139 (UL), N.A.S harp/NOAO/AURA/NSF; 139 (UR), ESO/L. Calçada; 139 (MC), NASA/ESA/Hubble Heritage Team (AURA/STScI); 139 (LL), R. Hynes/Louisiana State University; 140 (UL), NASA/ESA/Hubble Heritage Team (STScI/ARA); 140 (LL), NASA/ESA/Raghvendra Sahai/John Trauger (JPL)/WFPC2 science team; 140 (LR), ESO; 141 (UR), WikiSky; 141 (LL), NASA/ESA/Hubble Heritage Team (STScI/AURA); 141 (LC), ESO; 142, Wil Tirion; 143 (MC), Wikimedia Commons; 143 (LR), ESO/B. Tafreshi (twanight. org); 144 (UL), NASA/JPL-Caltech; 144 (UR), Casey Reed; 144 (MC), NASA/JPL; 144 (LL), NASA/ESA/R. Sankrit, W. Blair (Johns Hopkins University); 144–145, ESA/Hubble & NASA; 146 (UL), NASA/JPL/USGS; 146 (UR), NASA/JPL-Caltech/Harvard-Smithsonian CfA; 146 (ML), ESO; 146 (LL), X-ray: NASA/CXC/SAO/E. Nardini et al.; Optical: NASA/STScI; 146 (LC): ESO/L. Calçada; 147 (LL), NASA/JPL-Caltech; 148 (LL), Wikimedia Commons/Linda Hall Library; 149 (UL), ESO/J. Emerson/VISTA/Cambridge Astronomical Survey Unit; 149 (UR), NASA/ESA/M. Robberto (STScI/ESA)/Hubble Space Telescope Orion Treasury Project Team; 149 (LL), ESO/L. Calçada; 149 (LR), DESY/Science Communication Lab; 150 (LL), NASA/ESA/C.R. O'Dell (Rice University); 150 (LR), ESO/Igor Chekalin; 151 (ML), Wikimedia Commons/Rogelio Bernal Andreo; 151 (UR), ESO; 151 (LR), NASA/ESA/K. L. Luhman (Harvard-Smithsonian CfA)/G. Schneider, E. Young, G. Rieke, A. Cotera, H. Chen, M. Rieke, R. Thompson (Steward Observatory, University of Arizona); 152 (LL), ESO; 152 (LR), Rob van Gent (Utrecht University); 153 (UL), WikiSky; 153 (LL), ESO; 153 (LR), NASA/JPL; 154 (UR), ESA/Hubble & NASA; 154 (LL), NASA/ESA; 154 (LC), C. Marois et al. (NRC Canada); 154 (LR), Gemini Observatory/Travis Rector (University of Alaska Anchorage); 155 (UL), USNO; 155 (UR), NASA/ESA/Alfred Vidal-Madjar (Institut d'Astrophysique de Paris); 156 (LR), WikiSky; 157 (UL), ESA/Hubble & NASA; 157 (UC), ESO/M. Kornmesser/Nick Risinger (skysurvey.org); 157 (MC), ESA/Hubble & NASA/D. Milisavljevic (Purdue University); 157 (LL), Wikimedia Commons; 158 (UL), Wikimedia Commons; 158 (MR), NASA/ESA/Andy Fabian (University of Cambridge); 158 (LR), X-ray: NASA/CXC/RIKEN/D. Takei et al.; Optical: NASA/STScI; Radio:

INDEX

Illustrations are indicated in **BOLD**